LONDON MATHEMATICAL SOCIETY LECTURE NOTE SERIES

Managing Editor: Professor M. Reid, Mathematics Institute,
University of Warwick, Coventry CV4 7AL, United Kingdom

The titles below are available from booksellers, or from Cambridge University Press at http://www.cambridge.org/mathematics

Conference photograph

London Mathematical Society Lecture Note Series: 394

Variational Problems in Differential Geometry

University of Leeds 2009

Edited by

R. BIELAWSKI
K. HOUSTON
J.M. SPEIGHT
University of Leeds

 **CAMBRIDGE**
UNIVERSITY PRESS

CAMBRIDGE
UNIVERSITY PRESS

University Printing House, Cambridge CB2 8BS, United Kingdom

One Liberty Plaza, 20th Floor, New York, NY 10006, USA

477 Williamstown Road, Port Melbourne, VIC 3207, Australia

314-321, 3rd Floor, Plot 3, Splendor Forum, Jasola District Centre, New Delhi - 110025, India

103 Penang Road, #05-06/07, Visioncrest Commercial, Singapore 238467

Cambridge University Press is part of the University of Cambridge.

It furthers the University's mission by disseminating knowledge in the pursuit of education, learning and research at the highest international levels of excellence.

www.cambridge.org
Information on this title: www.cambridge.org/9780521282741

© Cambridge University Press 2012

First published 2012

A catalogue record for this publication is available from the British Library

Library of Congress Cataloging in Publication data
Variational problems in differential geometry : University of Leeds, 2009 /
edited by R. Bielawski, K. Houston, J.M. Speight.
p.   cm. – (London Mathematical Society lecture note series ; 394)
Includes bibliographical references.
ISBN 978-0-521-28274-1 (pbk.)
1. Geometry, Differential – Congresses.   I. Bielawski, R.   II. Houston, Kevin, 1968–
III. Speight, J. M. (J. Martin)   IV. Title. V. Series.
QA641.V37   2012
516.3′6 – dc23   2011027490

ISBN 978-0-521-28274-1 Paperback

Contents

9 Manifolds with k-positive Ricci curvature **182**

Jon Wolfson

Contributors

Bernd Ammann
Facultät für Mathematik, Universität Regensburg, 93040 Regensburg, Germany

Pierre Jammes
Laboratoire J.-A. Dieudonné, Université Nice – Sophia Antipolis, Parc Valrose, F-06108 NICE Cedex 02, France

Claudio Arezzo
Abdus Salam International Center for Theoretical Physics, Strada Costiera 11, Trieste (Italy) and Dipartimento di Matematica, Università di Parma, Parco Area delle Scienze 53/A, Parma, Italy

Alberto Della Vedova
Fine Hall, Princeton University, Princeton, NJ 08544 and Dipartimento di Matematica, Università di Parma, Parco Area delle Scienze 53/A, Parma, Italy

Gabriele La Nave
Department of Mathematics, Yeshiva University, 500 West 185 Street, New York, NY, USA

Paul Baird
Département de Mathématiques, Université de Bretagne Occidentale, 6 Avenue Le Gorgeu – CS 93837, 29238 Brest, France

Josef F. Dorfmeister
Fakultät für Mathematik, Technische Universität München, Boltzmannstr. 3, D-85747 Garching, Germany

Akito Futaki
*Department of Mathematics, Tokyo Institute of Technology, 2-12-1,
O-okayama, Meguro, Tokyo 152-8551, Japan*

Yuji Sano
*Department of Mathematics, Kyushu University, 6-10-1, Hakozaki,
Higashiku, Fukuoka-city, Fukuoka 812-8581 Japan*

Frédéric Hélein
*Institut de Mathématiques de Jussieu, UMR CNRS 7586, Université Denis
Diderot Paris 7, 175 rue du Chevaleret, 75013 Paris, France*

Lorenz J. Schwachhöfer
*Fakultät für Mathematik, Technische Universität Dortmund, Vogelpothsweg
87, 44221 Dortmund, Germany*

Richard A. Wentworth
*Department of Mathematics, University of Maryland, College Park, MD
20742, USA*

Graeme Wilkin
*Department of Mathematics, University of Colorado, Boulder, CO 80309,
USA*

Jon Wolfson
*Department of Mathematics, Michigan State University, East Lansing,
MI 48824, USA*

Preface

The workshop *Variational Problems in Differential Geometry* was held at the University of Leeds from March 30 to April 2nd, 2009.

The aim of the meeting was to bring together researchers working on disparate geometric problems, all of which admit a variational formulation. Among the topics discussed were recent developments in harmonic maps and morphisms, minimal and CMC surfaces, extremal Kähler metrics, the Yamabe functional, Hamiltonian variational problems, and topics related to gauge theory and to the Ricci flow.

The meeting incorporated a special session in honour of John C. Wood, on the occasion of his 60th birthday, to celebrate his seminal contributions to the theory of harmonic maps and morphisms.

The following mathematicians gave one-hour talks: Bernd Ammann, Claudio Arezzo, Paul Baird, Olivier Biquard, Christoph Boehm, Francis Burstall, Josef Dorfmeister, Akito Futaki, Mark Haskins, Frederic Helein, Nicolaos Kapouleas, Mario Micallef, Frank Pacard, Simon Salamon, Lorenz Schwachhoefer, Peter Topping, Richard Wentworth, and Jon Wolfson.

There were about 50 participants from the UK, US, Japan and several European countries. The schedule allowed plenty of opportunities for discussion and interaction between official talks and made for a successful and stimulating meeting.

The workshop was financially supported by the London Mathematical Society, the Engineering and Physical Sciences Research Council of Great Britain and the School of Mathematics, University of Leeds.

The articles presented in this volume represent the whole spectrum of the subject.

The supremum of first eigenvalues of conformally covariant operators in a conformal class by Ammann and Jammes is concerned with the first eigenvalues of the Yamabe operator, the Dirac operator, and more general conformally covariant elliptic operators on compact Riemannian manifolds. It is well known that the infimum of the first eigenvalue in a given conformal class reflects a rich geometric structure. In this article, the authors study the supremum of the first eigenvalue and show that, for a very general class of operators, this supremum is infinite.

The article, *K-Destabilizing test configurations with smooth central fiber* by Arezzo, Della Vedova, and La Nave is concerned with the famous Tian-Yau-Donaldson conjecture about existence of constant scalar curvature Kähler metrics. They construct many new families of *K*-unstable manifolds, and, consequently, many new examples of manifolds which do not admit Kähler constant scalar curvature metrics in some cohomology classes.

As has been now understood, a very natural extension of Einstein metrics are the Ricci solitons. These are the subject of Paul Baird's article *Explicit constructions of Ricci solitons*, in which he does precisely that: he constructs many explicit examples, including some in the more exotic geometries Sol_3, Nil_3, and Nil_4.

Josef Dorfmeister is concerned with a more classical topic: that of constant mean curvature and Willmore surfaces. In recent years, many new examples of such surfaces were constructed using loop groups. The method relies on finding "Iwasawa-like" decompositions of loop groups and the article *Open Iwasawa cells in twisted loop groups and some applications to harmonic maps* discusses such decompositions and their singularities.

The currently extremely important notions of *K*-stability and *K*-polystability are the topic of the paper by Futaki and Sano *Multiplier ideal sheaves and geometric problems*. This is an expository article giving state-of-the-art presentation of the powerful method of multiplier ideal sheaves and their applications to Kähler-Einstein and Sasaki-Einstein geometries.

Multisymplectic formalism and the covariant phase space by Frédéric Hélein takes us outside Riemannian geometry. The author presents an alternative (in fact, two of them) to the Feynman integral as a foundation of quantum field theory.

Lorenz Schwachhöfer's *Nonnegative curvature on disk bundles* is a survey of the glueing method used to construct Riemannian manifolds with nonnegative sectional curvature - one of the classical problems in geometry.

Morse theory and stable pairs by Wentworth and Wilkin introduces new techniques to compute equivariant cohomology of certain natural moduli spaces. The main ingredient is a version of Morse-Atiyah-Bott theory adapted to singular infinite dimensional spaces.

The final article, *Manifolds with k-positive Ricci curvature*, by Jon Wolfson, is a survey of results and conjectures about Riemannian *n*-manifolds with

k-positive Ricci curvature. These interpolate between positive scalar curvature (n-positive Ricci curvature) and positive Ricci curvature (1-positive Ricci curvature), and the author shows how the results about k-positive Ricci curvature, $1 < k < n$, also interpolate, or should do, between what is known about manifolds satisfying those two classical notions of positivity.

We would like to extend our thanks to our colleague John Wood for his help and assistance in preparing these proceedings.

R. Bielawski
K. Houston
J.M. Speight
Leeds, UK

1

The supremum of first eigenvalues of conformally covariant operators in a conformal class

BERND AMMANN AND PIERRE JAMMES

Abstract

Let (M, g) be a compact Riemannian manifold of dimension ≥ 3. We show that there is a metric $\tilde{g}$ conformal to g and of volume 1 such that the first positive eigenvalue of the conformal Laplacian with respect to $\tilde{g}$ is arbitrarily large. A similar statement is proven for the first positive eigenvalue of the Dirac operator on a spin manifold of dimension ≥ 2.

1.1 Introduction

The goal of this article is to prove the following theorems.

Theorem 1.1.1 *Let (M, g_0, χ) be compact Riemannian spin manifold of dimension $n \geq 2$. For any metric g in the conformal class $[g_0]$, we denote the first positive eigenvalue of the Dirac operator on (M, g, χ) by $\lambda_1^+(D_g)$. Then*

$$\sup_{g \in [g_0]} \lambda_1^+(D_g)\mathrm{Vol}(M, g)^{1/n} = \infty.$$

Theorem 1.1.2 *Let (M, g_0, χ) be compact Riemannian manifold of dimension $n \geq 3$. For any metric g in the conformal class $[g_0]$, we denote the first positive eigenvalue of the conformal Laplacian $L_g := \Delta_g + \frac{n-2}{4(n-1)}\mathrm{Scal}_g$ (also called Yamabe operator) on (M, g, χ) by $\lambda_1^+(L_g)$. Then*

$$\sup_{g \in [g_0]} \lambda_1^+(L_g)\mathrm{Vol}(M, g)^{2/n} = \infty.$$

The Dirac operator and the conformal Laplacian belong to a large family of operators, defined in details in subsection 1.2.3. These operators are

1

called conformally covariant elliptic operators of order k and of bidegree $((n-k)/2, (n+k)/2)$, acting on manifolds (M, g) of dimension $n > k$. In our definition we also claim formal self-adjointness.

All such conformally covariant elliptic operators of order k and of bidegree $((n-k)/2, (n+k)/2)$ share several analytical properties, in particular they are associated to the non-compact embedding $H^{k/2} \to L^{2n/(n-k)}$. Often they have interpretations in conformal geometry. To give an example, we define for a compact Riemannian manifold (M, g_0)

$$Y(M, [g_0]) := \inf_{g \in [g_0]} \lambda_1(L_g) \mathrm{Vol}(M, g)^{2/n},$$

where $\lambda_1(L_g)$ is the lowest eigenvalue of L_g. If $Y(M, [g_0]) > 0$, then the solution of the Yamabe problem [29] tells us that the infimum is attained and the minimizer is a metric of constant scalar curvature. This famous problem was finally solved by Schoen and Yau using the positive mass theorem.

In a similar way, for $n = 2$ the Dirac operator is associated to constant-mean-curvature conformal immersions of the universal covering into $\mathbb{R}^3$. If a Dirac-operator-analogue of the positive mass theorem holds for a given manifold (M, g_0), then the infimum

$$\inf_{g \in [g_0]} \lambda_1^+(D_g) \mathrm{Vol}(M, g)^{1/n}$$

is attained [3]. However, it is still unclear whether such a Dirac-operator-analogue of the positive mass theorem holds in general.

The Yamabe problem and its Dirac operator analogue, as well as the analogues for other conformally covariant operators are typically solved by minimizing an associated variational problem. As the Sobolev embedding $H^{k/2} \to L^{2n/(n-k)}$ is non-compact, the direct method of the calculus of variation fails, but perturbation techniques and conformal blow-up techniques typically work. Hence all these operators share many properties.

However, only few statements can be proven simultaneously for all conformally covariant elliptic operators of order k and of bidegree $((n-k)/2, (n+k)/2)$. Some of the operators are bounded from below (e.g. the Yamabe and the Paneitz operator), whereas others are not (e.g. the Dirac operator). Some of them admit a maximum principle, others do not. Some of them act on functions, others on sections of vector bundles. The associated Sobolev space $H^{k/2}$ has non-integer order if k is odd, hence it is not the natural domain of a differential operator. For Dirac operators, the spin structure has to be considered in order to derive a statement as Theorem 1.1.1 for $n = 2$. Because of these differences, most analytical properties have to be proven for each operator separately.

We consider it hence as remarkable that the proof of our Theorems 1.1.1 and 1.1.2 can be extended to all such operators. Our proof only uses some few properties of the operators, defined axiomatically in 1.2.3. More exactly we prove the following.

Theorem 1.1.3 *Let P_g be a conformally covariant elliptic operator of order k, of bidegree $((n - k)/2, (n + k)/2)$ acting on manifolds of dimension $n > k$. We also assume that P_g is invertible on $\mathbb{S}^{n-1} \times \mathbb{R}$ (see Definition 1.2.4). Let (M, g_0) be compact Riemannian manifold. In the case that P_g depends on the spin structure, we assume that M is oriented and is equipped with a spin structure. For any metric g in the conformal class $[g_0]$, we denote the first positive eigenvalue of P_g by $\lambda_1^+(P_g)$. Then*

$$\sup_{g \in [g_0]} \lambda_1^+(P_g) \mathrm{Vol}(M, g)^{k/n} = \infty.$$

The interest in this result is motivated by three questions. At first, as already mentioned above the infimum

$$\inf_{g \in [g_0]} \lambda_1^+(D_g) \mathrm{Vol}(M, g)^{1/n}$$

reflects a rich geometrical structure [3], [4], [5], [7], [8], similarly for the conformal Laplacian. It seems natural to study the supremum as well.

The second motivation comes from comparing Theorem 1.1.3 to results about some other differential operators. For the Hodge Laplacian Δ_p^g acting on p-forms, we have $\sup_{g \in [g_0]} \lambda_1(\Delta_p^g) \mathrm{Vol}(M, g)^{2/n} = +\infty$ for $n \geq 4$ and $2 \leq p \leq n - 2$ ([19]). On the other hand, for the Laplacian Δ^g acting on functions, we have

$$\sup_{g \in [g_0]} \lambda_k(\Delta^g) \mathrm{Vol}(M, g)^{2/n} < +\infty$$

(the case $k = 1$ is proven in [20] and the general case in [27]). See [25] for a synthetic presentation of this subject.

The essential idea in our proof is to construct metrics with longer and longer cylindrical parts. We will call this an *asymptotically cylindrical blowup*. Such metrics are also called *Pinocchio metrics* in [2, 6]. In [2, 6] the behavior of Dirac eigenvalues on such metrics has already been studied partially, but the present article has much stronger results. To extend these existing results provides the third motivation.

Acknowledgments We thank B. Colbois, M. Dahl, and E. Humbert for many related discussions. We thank R. Gover for some helpful comments on conformally covariant operators, and for several references. The first author wants to thank the Albert Einstein institute at Potsdam-Golm for its very kind hospitality which enabled to write the article.

1.2 Preliminaries

1.2.1 Notations

In this article $B_y(r)$ denotes the ball of radius r around y, $S_y(r) = \partial B_y(r)$ its boundary. The standard sphere $S_0(1) \subset \mathbb{R}^n$ in $\mathbb{R}^n$ is denoted by $\mathbb{S}^{n-1}$, its volume is ω_{n-1}. For the volume element of (M, g) we use the notation dv^g. In our article, $\Gamma(V)$ (resp. $\Gamma_c(V)$) always denotes the set of all smooth sections (resp. all compactly supported smooth sections) of the vector bundle $V \to M$.

For sections u of $V \to M$ over a Riemannian manifold (M, g) the Sobolev norms L^2 and H^s, $s \in \mathbb{N}$, are defined as

$$\|u\|^2_{L^2(M,g)} := \int_M |u|^2 \, dv^g$$
$$\|u\|^2_{H^s(M,g)} := \|u\|^2_{L^2(M,g)} + \|\nabla u\|^2_{L^2(M,g)} + \cdots + \|\nabla^s u\|^2_{L^2(M,g)}.$$

The vector bundle V will be suppressed in the notation. If M and g are clear from the context, we write just L^2 and H^s. The completion of $\{u \in \Gamma(V) \mid \|u\|_{H^s(M,g)} < \infty\}$ with respect to the $H^s(M, g)$-norm is denoted by $\Gamma_{H^s(M,g)}(V)$, or if (M, g) or V is clear from the context, we alternatively write $\Gamma_{H^s}(V)$ or $H^s(M, g)$ for $\Gamma_{H^s(M,g)}(V)$. The same definitions are used for L^2 instead of H^s. And similarly $\Gamma_{C^k(M,g)}(V) = \Gamma_{C^k}(V) = C^k(M, g)$ is the set of all C^k-sections, $k \in \mathbb{N} \cup \{\infty\}$.

1.2.2 Removal of singularities

In the proof we will use the following removal of singularities lemma.

Lemma 1.2.1 (Removal of singularities lemma) *Let Ω be a bounded open subset of $\mathbb{R}^n$ containing 0. Let P be an elliptic differential operator of order k on Ω, $f \in C^\infty(\Omega)$, and let $u \in C^\infty(\Omega \setminus \{0\})$ be a solution of*

$$Pu = f \tag{1.1}$$

on $\Omega \setminus \{0\}$ with

$$\lim_{\varepsilon \to 0} \int_{B_0(2\varepsilon) - B_0(\varepsilon)} |u| r^{-k} = 0 \quad and \quad \lim_{\varepsilon \to 0} \int_{B_0(\varepsilon)} |u| = 0 \tag{1.2}$$

where r is the distance to 0. Then u is a (strong) solution of (1.1) on Ω. The same result holds for sections of vector bundles over relatively compact open subset of Riemannian manifolds.

Proof We show that u is a weak solution of (1.1) in the distributional sense, and then it follows from standard regularity theory, that it is also a strong solution. This means that we have to show that for any given compactly supported smooth test function $\psi : \Omega \to \mathbb{R}$ we have

$$\int_\Omega u P^* \psi = \int_\Omega f\psi.$$

Let $\eta : \Omega \to [0, 1]$ be a test function that is identically 1 on $B_0(\varepsilon)$, has support in $B_0(2\varepsilon)$, and with $|\nabla^m \eta| \leq C_m/\varepsilon^m$. It follows that

$$\sup |P^*(\eta\psi)| \leq C(P, \Omega, \psi)\varepsilon^{-k},$$

on $B_0(2\varepsilon) \setminus B_0(\varepsilon)$ and $\sup |P^*(\eta\psi)| \leq C(P, \Omega, \psi)$ on $B_0(\varepsilon)$ and hence

$$
\left| \int_\Omega u P^*(\eta\psi) \right| \leq C\varepsilon^{-k} \int_{B_0(2\varepsilon)\setminus B_0(\varepsilon)} |u| + C \int_{B_0(\varepsilon)} |u|
$$
$$
\leq C \int_{B_0(2\varepsilon)\setminus B_0(\varepsilon)} |u|r^{-k} + C \int_{B_0(\varepsilon)} |u| \to 0.
$$
(1.3)

We conclude

$$
\int_\Omega u P^*\psi = \int_\Omega u P^*(\eta\psi) + \int_\Omega u P^*((1 - \eta)\psi)
$$
$$
= \underbrace{\int_\Omega u P^*(\eta\psi)}_{\to 0} + \underbrace{\int_\Omega (Pu)(1 - \eta)\psi}_{\to \int_\Omega f\psi}
$$
(1.4)

for $\varepsilon \to 0$. Hence the lemma follows. $\qquad\square$

Condition (1.2) is obviously satisfied if $\int_\Omega |u|r^{-k} < \infty$. It is also satisfied if

$$\int_\Omega |u|^2 r^{-k} < \infty \text{ and } k \leq n,$$
(1.5)

as in this case

$$
\left(\int_{B_0(2\varepsilon)\setminus B_0(\varepsilon)} |u|r^{-k} \right)^2 \leq \int_\Omega |u|^2 r^{-k} \underbrace{\int_{B_0(2\varepsilon)\setminus B_0(\varepsilon)} r^{-k}}_{\leq C}.
$$

1.2.3 Conformally covariant elliptic operators

In this subsection we present a class of certain conformally covariant elliptic operators. Many important geometric operators are in this class, in particular the conformal Laplacian, the Paneitz operator, the Dirac operator, see also [18, 21, 22] for more examples. Readers who are only interested in the Dirac operator, the Conformal Laplacian or the Paneitz operator, can skip this part and continue with section 1.3.

Such a conformally covariant operator is not just one single differential operator, but a procedure how to associate to an n-dimensional Riemannian manifold (M, g) (potentially with some additional structure) a differential operator P_g of order k acting on a vector bundle. The important fact is that if $g_2 = f^2 g_1$, then one claims

$$P_{g_2} = f^{-\frac{n+k}{2}} P_{g_1} f^{\frac{n-k}{2}}. \tag{1.6}$$

One also expresses this by saying that P has bidegree $((n - k)/2, (n + k)/2)$.

The sense of this equation is apparent if P_g is an operator from $C^\infty(M)$ to $C^\infty(M)$. If P_g acts on a vector bundle or if some additional structure (as e.g. spin structure) is used for defining it, then a rigorous and careful definition needs more attention. The language of categories provides a good formal framework [30]. The concept of conformally covariant elliptic operators is already used by many authors, but we do not know of a reference where a formal definition is carried out that fits to our context. (See [26] for a similar categorial approach that includes some of the operators presented here.) Often an intuitive definition is used. The intuitive definition is obviously sufficient if one deals with operators acting on functions, such as the conformal Laplacian or the Paneitz operator. However to properly state Theorem 1.1.3 we need the following definition.

Let $Riem^n$ (resp. $Riemspin^n$) be the category n-dimensional Riemannian manifolds (resp. n-dimensional Riemannian manifolds with orientation and spin structure). Morphisms from (M_1, g_1) to (M_2, g_2) are conformal embeddings $(M_1, g_1) \hookrightarrow (M_2, g_2)$ (resp. conformal embeddings preserving orientation and spin structure).

Let $Laplace_k^n$ (resp. $Dirac_k^n$) be the category whose objects are

$$\{(M, g), V_g, P_g\}$$

where (M, g) in an object of $Riem^n$ (resp. $Riemspin^n$), where V_g is a vector bundle with a scalar product on the fibers, where $P_g : \Gamma(V_g) \to \Gamma(V_g)$ is an elliptic formally self-adjoint differential operator of order k.

A morphism (ι, κ) from $\{(M_1, g_1), V_{g_1}, P_{g_1}\}$ to $\{(M_2, g_2), V_{g_2}, P_{g_2}\}$ consists of a conformal embedding $\iota : (M_1, g_1) \hookrightarrow (M_2, g_2)$ (preserving orientation and spin structure in the case of $Dirac_k^n$) together with a fiber isomorphism $\kappa : \iota^* V_{g_2} \to V_{g_1}$ preserving fiberwise length, such that P_{g_1} and P_{g_2} satisfy the conformal covariance property (1.6). For stating this property precisely, let $f > 0$ be defined by $\iota^* g_2 = f^2 g_1$, and let $\kappa_* : \Gamma(V_{g_2}) \to \Gamma(V_{g_1})$, $\kappa_*(\varphi) = \kappa \circ \varphi \circ \iota$. Then the conformal covariance property is

$$\kappa_* P_{g_2} = f^{-\frac{n+k}{2}} P_{g_1} f^{\frac{n-k}{2}} \kappa_*. \tag{1.7}$$

In the following the maps κ and ι will often be evident from the context and then will be omitted. The transformation formula (1.7) then simplifies to (1.6).

Definition 1.2.2 A *conformally covariant elliptic operator of order k and of bidegree $((n-k)/2, (n+k)/2)$* is a contravariant functor from $Riem^n$ (resp. $Riemspin^n$) to $Laplace_k^n$ (resp. $Dirac_k^n$), mapping (M, g) to (M, g, V_g, P_g) in such a way that the coefficients are continuous in the C^k-topology of metrics (see below). To shorten notation, we just write P_g or P for this functor.

It remains to explain the C^k-continuity of the coefficients.

For Riemannian metrics g, g_1, g_2 defined on a compact set $K \subset M$ we set

$$d_{C^k(K)}^g(g_1, g_2) := \max_{t=0,\dots,k} \|(\nabla_g)^t(g_1 - g_2)\|_{C^0(K)}.$$

For a fixed background metric g, the relation $d_{C^k(K)}^g(\cdot, \cdot)$ defines a distance function on the space of metrics on K. The topology induced by d^g is independent of this background metric and it is called the C^k-*topology of metrics on K*.

Definition 1.2.3 We say that the coefficients of P are *continuous in the C^k-topology of metrics* if for any metric g on a manifold M, and for any compact subset $K \subset M$ there is a neighborhood $\mathcal{U}$ of $g|_K$ in the C^k-topology of metrics on K, such that for all metrics $\tilde{g}$, $\tilde{g}|_K \in \mathcal{U}$, there is an isomorphism of vector bundles $\hat{\kappa} : V_g|_K \to V_{\tilde{g}}|_K$ over the identity of K with induced map $\hat{\kappa}_* : \Gamma(V_g|_K) \to \Gamma(V_{\tilde{g}}|_K)$ with the property that the coefficients of the differential operator

$$P_g - (\hat{\kappa}_*)^{-1} P_{\tilde{g}} \hat{\kappa}_*$$

depend continuously on $\tilde{g}$ (with respect to the C^k-topology of metrics).

1.2.4 Invertibility on $\mathbb{S}^{n-1} \times \mathbb{R}$

Let P be a conformally covariant elliptic operator of order k and of bidegree $((n - k)/2, (n + k)/2)$. For $(M, g) = \mathbb{S}^{n-1} \times \mathbb{R}$, the operator P_g is a self-adjoint operator $H^k \subset L^2 \to L^2$ (see Lemma 1.3.1 and the comments thereafter).

Definition 1.2.4 We say that *P is invertible on* $\mathbb{S}^{n-1} \times \mathbb{R}$ if P_g is an invertible operator $H^k \to L^2$ where g is the standard product metric on $\mathbb{S}^{n-1} \times \mathbb{R}$. In order words there is a constant $\sigma > 0$ such that the spectrum of $P_g : \Gamma_{H^k}(V_g) \to \Gamma_{L^2}(V_g)$ is contained in $(-\infty, -\sigma] \cup [\sigma, \infty)$ for any $g \in U$. In the following, the largest such σ will be called σ_P.

We conjecture that any conformally covariant elliptic operator of order k and of bidegree $((n - k)/2, (n + k)/2)$ with $k < n$ is invertible on $\mathbb{S}^{n-1} \times \mathbb{R}$.

1.2.5 Examples

Example 1: The Conformal Laplacian
Let

$$L_g := \Delta_g + \frac{n - 2}{4(n - 1)} \mathrm{Scal}_g,$$

be the conformal Laplacian. It acts on functions on a Riemannian manifold (M, g), i.e. V_g is the trivial real line bundle $\mathbb{R}$. Let $\iota : (M_1, g_1) \hookrightarrow (M_2, g_2)$ be a conformal embedding. Then we can choose $\kappa := \mathrm{Id} : \iota^* V_{g_2} \to V_{g_1}$ and formula (1.7) holds for $k = 2$ (see e.g. [15, Section 1.J]). All coefficients of L_g depend continuously on g in the C^2-topology. Hence L is a conformally covariant elliptic operator of order 2 and of bidegree $((n - 2)/2, (n + 2)/2)$.

The scalar curvature of $\mathbb{S}^{n-1} \times \mathbb{R}$ is $(n - 1)(n - 2)$. The spectrum of L_g on $\mathbb{S}^{n-1} \times \mathbb{R}$ of L_g coincides with the essential spectrum of L_g and is $[\sigma_L, \infty)$ with $\sigma_L := (n - 2)^2/4$. Hence L is invertible on $S^{n-1} \times \mathbb{R}$ if (and only if) $n > 2$.

Example 2: The Paneitz operator
Let (M, g) be a smooth, compact Riemannian manifold of dimension $n \geq 5$. The Paneitz operator P_g is given by

$$P_g u = (\Delta_g)^2 u - \mathrm{div}_g(A_g \, du) + \frac{n - 4}{2} Q_g u$$

where

$$A_g := \frac{(n - 2)^2 + 4}{2(n - 1)(n - 2)} \mathrm{Scal}_g \, g - \frac{4}{n - 2} \mathrm{Ric}_g,$$

$$Q_g = \frac{1}{2(n - 1)} \Delta_g \mathrm{Scal}_g + \frac{n^3 - 4n^2 + 16n - 16}{8(n - 1)^2(n - 2)^2} \mathrm{Scal}_g^2 - \frac{2}{(n - 2)^2} |\mathrm{Ric}_g|^2.$$

This operator was defined by Paneitz [32] in the case $n = 4$, and it was generalized by Branson in [17] to arbitrary dimensions ≥ 4. We also refer to Theorem 1.21 of the overview article [16]. The explicit formula presented above can be found e.g. in [23]. The coefficients of P_g depend continuously on g in the C^4-topology

As in the previous example we can choose for κ the identity, and then the Paneitz operator P_g is a conformally covariant elliptic operator of order 4 and of bidegree $((n-4)/2, (n+4)/2)$.

On $\mathbb{S}^{n-1} \times \mathbb{R}$ one calculates

$$A_g = \frac{(n-4)n}{2} \mathrm{Id} + 4\pi_{\mathbb{R}} > 0$$

where $\pi_{\mathbb{R}}$ is the projection to vectors parallel to $\mathbb{R}$.

$$Q_g = \frac{(n-4)n^2}{8}.$$

We conclude

$$\sigma_P = Q = \frac{(n-4)n^2}{8}$$

and P is invertible on $\mathbb{S}^{n-1} \times \mathbb{R}$ if (and only if) $n > 4$.

Examples 3: The Dirac operator.

Let $\tilde{g} = f^2 g$. Let $\Sigma_g M$ resp. $\Sigma_{\tilde{g}} M$ be the spinor bundle of (M, g) resp. $(M, \tilde{g})$. Then there is a fiberwise isomorphism $\beta_{\tilde{g}}^g : \Sigma_g M \to \Sigma_{\tilde{g}} M$, preserving the norm such that

$$D_{\tilde{g}} \circ \beta_{\tilde{g}}^g(\varphi) = f^{-\frac{n+1}{2}} \beta_{\tilde{g}}^g \circ D_g \left(f^{\frac{n-1}{2}} \varphi \right),$$

see [24, 14] for details. Furthermore, the cocycle conditions

$$\beta_{\tilde{g}}^g \circ \beta_g^{\tilde{g}} = \mathrm{Id} \qquad \text{and} \qquad \beta_{\hat{g}}^g \circ \beta_{\hat{g}}^{\tilde{g}} \circ \beta_{\tilde{g}}^g = \mathrm{Id}$$

hold for conformal metrics g, $\tilde{g}$ and $\hat{g}$. We will hence use the map $\beta_{\tilde{g}}^g$ to identify $\Sigma_g M$ with $\Sigma_{\tilde{g}} M$. Hence we simply get

$$D_{\tilde{g}} \varphi = f^{-\frac{n+1}{2}} \circ D_g \left(f^{\frac{n-1}{2}} \varphi \right). \tag{1.8}$$

All coefficients of D_g depend continuously on g in the C^1 topology. Hence D is a conformally covariant elliptic operator of order 1 and of bidegree $((n-1)/2, (n+1)/2)$.

The Dirac operator on $\mathbb{S}^{n-1} \times \mathbb{R}$ can be decomposed in a part D_{vert} deriving along $\mathbb{S}^{n-1}$ and a part D_{hor} deriving along $\mathbb{R}$, $D_g = D_{\mathrm{vert}} + D_{\mathrm{hor}}$, see [1] or [2].

Locally

$$D_{\text{vert}} = \sum_{i=1}^{n-1} e_i \cdot \nabla_{e_i}$$

for a local frame $(e_1, \ldots, e_{n-1})$ of $\mathbb{S}^{n-1}$. Here $\cdot$ denotes the Clifford multiplication $TM \otimes \Sigma_g M \to \Sigma_g M$. Furthermore $D_{\text{hor}} = \partial_t \cdot \nabla_{\partial_t}$, where $t \in \mathbb{R}$ is the standard coordinate of $\mathbb{R}$. The operators D_{vert} and D_{hor} anticommute. For $n \geq 3$, the spectrum of D_{vert} coincides with the spectrum of the Dirac operator on $\mathbb{S}^{n-1}$, we cite [12] and obtain

$$\text{spec} D_{\text{vert}} = \left\{ \pm \left(\frac{n-1}{2} + k \right) \mid k \in \mathbb{N}_0 \right\}.$$

The operator $(D_{\text{hor}})^2$ is the ordinary Laplacian on $\mathbb{R}$ and hence has spectrum $[0, \infty)$. Together this implies that the spectrum of the Dirac operator on $\mathbb{S}^{n-1} \times \mathbb{R}$ is the set $(-\infty, -\sigma_D] \cup [\sigma_D, \infty)$ with $\sigma_D = \frac{n-1}{2}$.

In the case $n = 2$ these statements are only correct if the circle $\mathbb{S}^{n-1} = \mathbb{S}^1$ carries the spin structure induced from the ball. Only this spin structure extends to the conformal compactification that is given by adding one point at infinity for each end. For this reason, we will understand in the whole article that all circles $\mathbb{S}^1$ should be equipped with this bounding spin structure. The extension of the spin structure is essential in order to have a spinor bundle on the compactification. The methods used in our proof use this extension implicitly.

Hence D is invertible on $S^{n-1} \times \mathbb{R}$ if (and only if) $n > 1$.

Most techniques used in the literature on estimating eigenvalues of the Dirac operators do not use the spin structure and hence these techniques cannot provide a proof in the case $n = 2$.

Example 4: The Rarita-Schwinger operator and many other Fegan type operators are conformally covariant elliptic operators of order 1 and of bidegree $((n-1)/2, (n+1)/2)$. See [21] and in the work of T. Branson for more information.

Example 5: Assume that (M, g) is a Riemannian spin manifold that carries a vector bundle $W \to M$ with metric and metric connection. Then there is a natural first order operator $\Gamma(\Sigma_g M \otimes W) \to \Gamma(\Sigma_g M \otimes W)$, the *Dirac operator twisted by W*. This operator has similar properties as conformally covariant elliptic operators of order 1 and of bidegree $((n-1)/2, (n+1)/2)$. The methods of our article can be easily adapted in order to show that Theorem 1.1.3 is also true for this twisted Dirac operator. However, twisted Dirac operators are not "conformally covariant elliptic operators" in the above sense. They could have been included in this class by replacing the category *Riemspin^n* by

Figure 1.1 Asymptotically cylindrical metrics g_L (alias Pinocchio metrics) with
growing nose length L.

a category of Riemannian spin manifolds with twisting bundles. In order not to
overload the formalism we chose not to present these larger categories.

The same discussion applies to the spin^c-Dirac operator of a spin^c-manifold.

1.3 Asymptotically cylindrical blowups

1.3.1 Convention

*From now on we suppose that P_g is a conformally covariant elliptic operator of
order k, of bidegree $((n - k)/2, (n + k)/2)$, acting on manifolds of dimension
n and invertible on $\mathbb{S}^{n-1} \times \mathbb{R}$.*

1.3.2 Definition of the metrics

Let g_0 be a Riemannian metric on a compact manifold M. We can suppose
that the injectivity radius in a fixed point $y \in M$ is larger than 1. The geodesic
distance from y to x is denoted by $d(x, y)$.

We choose a smooth function $F_\infty : M \setminus \{y\} \to [1, \infty)$ such such that
$F_\infty(x) = 1$ if $d(x, y) \geq 1$, $F_\infty(x) \leq 2$ if $d(x, y) \geq 1/2$ and such that $F_\infty(x) =
d(x, y)^{-1}$ if $d(x, y) \in (0, 1/2]$. Then for $L \geq 1$ we define F_L to be a smooth
positive function on M, depending only on $d(x, y)$, such that $F_L(x) = F_\infty(x)$
if $d(x, y) \geq e^{-L}$ and $F_L(x) \leq d(x, y)^{-1} = F_\infty(x)$ if $d(x, y) \leq e^{-L}$.

For any $L \geq 1$ or $L = \infty$ set $g_L := F_L^2 g_0$. The metric g_∞ is a complete
metric on M_∞.

The family of metrics (g_L) is called an *asymptotically cylindrical blowup*,
in the literature it is denoted as a family of *Pinocchio metrics* [6], see also
Figure 1.1.

1.3.3 Eigenvalues and basic properties on (M, g_L)

For the P-operator associated to (M, g_L), $L \in \{0\} \cup [1, \infty)$ (or more exactly
its self-adjoint extension) we simply write P_L instead of P_{g_L}. As M is compact
the spectrum of P_L is discrete.

We will denote the spectrum of P_L in the following way

$$\ldots \leq \lambda_1^-(P_L) < 0 = 0 \ldots = 0 < \lambda_1^+(P_L) \leq \lambda_2^+(P_L) \leq \ldots,$$

where each eigenvalue appears with the multiplicity corresponding to the dimension of the eigenspace. The zeros might appear on this list or not, depending on whether P_L is invertible or not. The spectrum might be entirely positive (for example the conformal Laplacian Y_g on the sphere) in which case $\lambda_1^-(P_L)$ is not defined. Similarly, $\lambda_1^+(P_L)$ is not defined if the spectrum of (P_L) is negative.

1.3.4 Analytical facts about (M_∞, g_∞)

The analysis of non-compact manifolds as (M_∞, g_∞) is more complicated than in the compact case. Nevertheless (M_∞, g_∞) is an asymptotically cylindrical manifold, and for such manifolds an extensive literature is available. One possible approach would be Melrose's b-calculus [31]: our cylindrical manifold is such a *b*-manifold, but for simplicity and self-containedness we avoid this theory. We will need some few properties that we will summarize in the following proposition.

We assume in the whole section that P is a conformally covariant elliptic operator that is invertible on $\mathbb{S}^{n-1} \times \mathbb{R}$, and we write $P_\infty := P_{g_\infty}$ for the operator acting on sections of the bundle V over (M_∞, g_∞).

Proposition 1.3.1 *P_∞ extends to a bounded operator from*

$$\Gamma_{H^k(M_\infty, g_\infty)}(V) \to \Gamma_{L^2(M_\infty, g_\infty)}(V)$$

and it satisfies the following regularity estimate

$$\|(\nabla^\infty)^s u\|_{L^2(M_\infty, g_\infty)} \leq C(\|u\|_{L^2(M_\infty, g_\infty)} + \|P_\infty u\|_{L^2(M_\infty, g_\infty)}) \qquad (1.9)$$

for all $u \in \Gamma_{H^k(M_\infty, g_\infty)}(V)$ and all $s \in \{0, 1, \ldots, k\}$. The operator

$$P_\infty : \Gamma_{H^k(M_\infty, g_\infty)}(V) \to \Gamma_{L^2(M_\infty, g_\infty)}(V)$$

is self-adjoint in the sense of an operator in $\Gamma_{L^2(M_\infty, g_\infty)}(V)$.

The proof of the proposition will be sketched in the appendix.

Proposition 1.3.2 *The essential spectrum of P_∞ coincides with the essential spectrum of the P-operator on the standard cylinder $\mathbb{S}^{n-1} \times \mathbb{R}$. Thus the essential spectrum of P_∞ is contained in $(-\infty, -\sigma_P] \cup [\sigma_P, \infty)$.*

This proposition follows from the characterization of the essential spectrum in terms of Weyl sequences, a well-known technique which is for example carried out and well explained in [13].

The second proposition states that the spectrum of P_∞ in the interval $(-\sigma_P, \sigma_P)$ is discrete as well. Eigenvalues of P_∞ in this interval will be called *small eigenvalues of* P_∞. Similarly to above we use the notation $\lambda_j^\pm(P_\infty)$ for the small eigenvalues of P_∞.

1.3.5 The kernel

Having recalled these well-known facts we will now study the kernel of conformally covariant operators.

If g and $\tilde{g} = f^2$ are conformal metrics on a compact manifold M, then

$$\varphi \mapsto f^{-\frac{n-k}{2}} \varphi$$

obviously defines an isomorphism from $\ker P_g$ to $\ker P_{\tilde{g}}$. It is less obvious that a similar statement holds if we compare g_0 and g_∞ defined before:

Proposition 1.3.3 *The map*

$$\ker P_0 \to \ker P_\infty$$

$$\varphi_0 \mapsto \varphi_\infty = F_\infty^{-\frac{n-k}{2}} \varphi_0$$

is an isomorphism of vector spaces.

Proof Suppose $\varphi_0 \in \ker P_0$. Using standard regularity results it is clear that $\sup |\varphi_0| < \infty$. Then

$$\int_{M_\infty} |\varphi_\infty|^2 \, dv^{g_\infty} \leq \int_{M \setminus B_y(1/2)} |\varphi_\infty|^2 \, dv^{g_\infty} + \sup |\varphi_0|^2 \int_{B_y(1/2)} F_\infty^{-(n-k)} \, dv^{g_\infty}$$

$$\leq 2^k \int_{M \setminus B_y(1/2)} |\varphi_0|^2 \, dv^{g_0} + \sup |\varphi_0|^2 \omega_{n-1} \int_0^{1/2} \frac{r^{n-1}}{r^k} \, dr < \infty.$$

$$(1.10)$$

Here we used that up to lower order terms dv^{g_∞} coincides with the product measure of the standard measure on the sphere with the measure $d(\log r) = \frac{1}{r} dr$. Furthermore, formula (1.6) implies $P_\infty \varphi_\infty = 0$. Hence the map is well-defined. In order to show that it is an isomorphism we show that the obvious inverse $\varphi_\infty \mapsto \varphi_0 := F_\infty^{\frac{n-k}{2}} \varphi_\infty$ is well defined. To see this we start with an L^2-section in the kernel of P_∞.

We calculate

$$\int_M F_\infty^k |\varphi_0|^2 \, dv^{g_0} = \int_{M_\infty} |\varphi_\infty|^2 \, dv^{g_\infty}.$$

Using again (1.6) we see that this section satisfies $P_0\varphi_0 = 0$ on $M \setminus \{y\}$. Hence condition (1.5) is satisfied, and together with the removal of singularity lemma (Lemma 1.2.1) one obtains that the inverse map is well defined. The proposition follows. □

1.4 Proof of the main theorem

1.4.1 Stronger version of the main theorem

We will now show the following theorem.

Theorem 1.4.1 *Let P be a conformally covariant elliptic operator of order k, of bidegree $((n-k)/2, (n+k)/2)$, on manifolds of dimension $n > k$. We assume that P is invertible on $\mathbb{S}^{n-1} \times \mathbb{R}$.*

If $\liminf_{L \to \infty} |\lambda_j^\pm(P_L)| < \sigma_P$, then

$$\lambda_j^\pm(P_L) \to \lambda_j^\pm(P_\infty) \in (-\sigma_P, \sigma_P) \qquad for \ L \to \infty.$$

In the case $\mathrm{Spec}(P_{g_0}) \subset (0, \infty)$ the theorem only makes a statement about λ_j^+, and conversely in the case that $\mathrm{Spec}(P_{g_0}) \subset (-\infty, 0)$ it only makes a statement about λ_j^-.

Obviously this theorem implies Theorem 1.1.3.

1.4.2 The supremum part of the proof of Theorem 1.4.1

At first we prove that

$$\limsup_{L \to \infty}(\lambda_j^+(P_L)) \le \lambda_j^+(P_\infty). \tag{1.11}$$

Let $\varphi_1, \ldots, \varphi_j$ be sequence of L^2-orthonormal eigenvectors of P_∞ to eigenvalues $\lambda_1^+(P_\infty), \ldots, \lambda_j^+(P_\infty) \in [-\bar\lambda, \bar\lambda]$, $\bar\lambda < \sigma_P$. We choose a cut-off function $\chi : M \to [0, 1]$ with $\chi(x) = 1$ for $-\log(d(x, y)) \le T$, $\chi(y) = 0$ for $-\log(d(x, y)) \ge 2T$, and $|(\nabla^\infty)^s \chi|_{g_\infty} \le C_s/T^s$ for all $s \in \{0, \ldots, k\}$.

Let φ be a linear combination of the eigenvectors $\varphi_1, \ldots, \varphi_j$. From Proposition 1.3.1 we see that

$$\|(\nabla^\infty)^s \varphi\|_{L^2(M_\infty, g_\infty)} \le C \|\varphi\|_{L^2(M_\infty, g_\infty)}$$

where C only depends on (M_∞, g_∞). Hence for sufficiently large T

$$\|P_\infty(\chi\varphi) - \chi P_\infty \varphi\|_{L^2(M_\infty, g_\infty)} \leq kC/T \|\varphi\|_{L^2(M_\infty, g_\infty)} \leq 2kC/T \|\chi\varphi\|_{L^2(M_\infty, g_\infty)}$$

as $\|\chi\varphi\|_{L^2(M_\infty, g_\infty)} \to \|\varphi\|_{L^2(M_\infty, g_\infty)}$ for $T \to \infty$. The section $\chi\varphi$ can be interpreted as a section on (M, g_L) if $L > 2T$, and on the support of $\chi\varphi$ we have $g_L = g_\infty$ and $P_\infty(\chi\varphi) = P_L(\chi\varphi)$. Hence standard Rayleigh quotient arguments imply that if P_∞ has m eigenvalues (counted with multiplicity) in the interval $[a, b]$ then P_L has m eigenvalues in the interval $[a - 2kC/T, b + 2kC/T]$. Taking the limit $T \to \infty$ we obtain (1.11).

By exchanging some obvious signs we obtain similarly

$$\limsup_{L \to \infty}(-\lambda_j^-(P_L)) \leq -\lambda_j^-(P_\infty). \tag{1.12}$$

1.4.3 The infimum part of the proof of Theorem 1.4.1

We now prove

$$\liminf_{L \to \infty}(\pm\lambda_j^\pm(P_L)) \geq \pm\lambda_j^\pm(P_\infty). \tag{1.13}$$

We assume that we have a sequence $L_i \to \infty$, and that for each i we have a system of orthogonal eigenvectors $\varphi_{i,1}, \ldots, \varphi_{i,m}$ of P_{L_i}, i.e. $P_{L_i}\varphi_{i,\ell} = \lambda_{i,\ell}\varphi_{i,\ell}$ for $\ell \in \{1, \ldots, m\}$. Furthermore we suppose that $\lambda_{i,\ell} \to \bar\lambda_\ell \in (-\sigma_P, \sigma_P)$ for $\ell \in \{1, \ldots, m\}$.

Then

$$\psi_{i,\ell} := \left(\frac{F_{L_i}}{F_\infty}\right)^{\frac{n-k}{2}} \varphi_{i,\ell}$$

satisfies

$$P_\infty \psi_{i,\ell} = h_{i,\ell}\psi_{i,\ell} \qquad \text{with} \qquad h_{i,\ell} := \left(\frac{F_{L_i}}{F_\infty}\right)^k \lambda_{i,\ell}.$$

Furthermore

$$\|\psi_{i,\ell}\|^2_{L^2(M_\infty, g_\infty)} = \int_M \left(\frac{F_{L_i}}{F_\infty}\right)^{-k} |\varphi_{i,\ell}|^2 \, dv^{g_{L_i}}$$

$$\leq \sup_M |\varphi_{i,\ell}|^2 \int_M \left(\frac{F_{L_i}}{F_\infty}\right)^{-k} dv^{g_{L_i}}$$

Because of

$$\int_M \left(\frac{F_{L_i}}{F_\infty}\right)^{-k} dv^{g_L} \leq C \int r^{n-1-k} \, dr < \infty$$

(for $n > k$) the norm $\|\psi_{i,\ell}\|_{L^2(M_\infty, g_\infty)}$ is finite as well, and we can renormalize such that

$$\|\psi_{i,\ell}\|_{L^2(M_\infty, g_\infty)} = 1.$$

Lemma 1.4.2 *For any $\delta > 0$ and any $\ell \in \{0, \ldots, m\}$ the sequence*

$$\left(\|\psi_{i,\ell}\|_{C^{k+1}(M \setminus B_y(\delta), g_\infty)}\right)_i$$

is bounded.

Proof of the lemma. After removing finitely many i, we can assume that $\lambda_i \leq 2\bar{\lambda}$ and $e^{-L_i} < \delta/2$. Hence $F_L = F_\infty$ and $h_i = \lambda_i$ on $M \setminus B_y(\delta/2)$. Because of

$$\int_{M \setminus B_y(\delta/2)} |(P_\infty)^s \psi_i|^2 \, dv^{g_\infty} \leq (2\bar{\lambda})^{2s} \int_{M \setminus B_y(\delta/2)} |\psi_i|^2 \, dv^{g_\infty} \leq (2\bar{\lambda})^{2s}$$

we obtain boundedness of ψ_i in the Sobolev space $H^{sk}(M \setminus B_y(3\delta/4), g_\infty)$, and hence, for sufficiently large s boundedness in $C^{k+1}(M \setminus B_y(\delta), g_\infty)$. The lemma is proved. $\qquad\square$

Hence after passing to a subsequence $\psi_{i,\ell}$ converges in $C^{k,\alpha}(M \setminus B_y(\delta), g_\infty)$ to a solution $\bar{\psi}_\ell$ of

$$P_\infty \bar{\psi}_\ell = \bar{\lambda}_\ell \bar{\psi}_\ell.$$

By taking a diagonal sequence, one can obtain convergence in $C_{\text{loc}}^{k,\alpha}(M_\infty)$ of $\psi_{i,\ell}$ to $\bar{\psi}_\ell$. It remains to prove that $\bar{\psi}_1, \ldots, \bar{\psi}_m$ are linearly independent, in particular that any $\bar{\psi}_\ell \neq 0$. For this we use the following lemma.

Lemma 1.4.3 *For any $\varepsilon > 0$ there is δ_0 and i_0 such that*

$$\|\psi_{i,\ell}\|_{L^2(B_y(\delta_0), g_\infty)} \leq \varepsilon \|\psi_{i,\ell}\|_{L^2(M_\infty, g_\infty)}$$

for all $i \geq i_0$ and all $\ell \in \{0, \ldots, m\}$. In particular,

$$\|\psi_{i,\ell}\|_{L^2(M \setminus B_y(\delta_0), g_\infty)} \geq (1 - \varepsilon) \|\psi_{i,\ell}\|_{L^2(M_\infty, g_\infty)}.$$

Proof of the lemma. Because of Proposition 1.3.1 and

$$\|P_\infty \psi_{i,\ell}\|_{L^2(M_\infty, g_\infty)} \leq |\bar{\lambda}_\ell| \, \|\psi_{i,\ell}\|_{L^2(M_\infty, g_\infty)} = |\bar{\lambda}_\ell|$$

we get

$$\|(\nabla^\infty)^s \psi_{i,\ell}\|_{L^2(M_\infty, g_\infty)} \leq C$$

for all $s \in \{0, \ldots, k\}$. Let χ be a cut-off function as in Subsection 1.4.2 with $T = -\log \delta$. Hence

$$\|P_\infty((1 - \chi)\psi_{i,\ell}) - (1 - \chi)P_\infty(\psi_{i,\ell})\|_{L^2(M_\infty, g_\infty)} \leq \frac{C}{T} = \frac{C}{-\log \delta}. \quad (1.14)$$

On the other hand $(B_y(\delta) \setminus \{y\}, g_\infty)$ converges for suitable choices of base points for $\delta \to 0$ to $\mathbb{S}^{n-1} \times (0, \infty)$ in the C^∞-topology of Riemannian manifolds with base points. Hence there is a function $\tau(\delta)$ converging to 0 such that

$$\|P_\infty((1 - \chi)\psi_{i,\ell})\|_{L^2(M_\infty, g_\infty)} \geq (\sigma_p - \tau(\delta))\|(1 - \chi)\psi_{i,\ell}\|_{L^2(M_\infty, g_\infty)}. \quad (1.15)$$

Using the obvious relation

$$\|(1 - \chi)P_\infty(\psi_{i,\ell})\|_{L^2(M_\infty, g_\infty)} \leq |\lambda_{i,\ell}| \, \|(1 - \chi)\psi_{i,\ell}\|_{L^2(M_\infty, g_\infty)}$$

we obtain with (1.14) and (1.15)

$$\|\psi_{i,\ell}\|_{L^2(B_y(\delta^2)), g_\infty)} \leq \|(1 - \chi)\psi_{i,\ell}\|_{L^2(M_\infty, g_\infty)}$$
$$\leq \frac{C}{|\log \delta|(\sigma_P - \tau(\delta) - |\lambda_{i,\ell}|)}.$$

The right hand side is smaller than ε for i sufficiently large and δ sufficiently small. The main statement of the lemma then follows for $\delta_0 := \delta^2$. The Minkowski inequality yields.

$$\|\psi_{i,\ell}\|_{L^2(M \setminus B_y(\delta^2)), g_\infty)} \geq 1 - \|\psi_{i,\ell}\|_{L^2(B_y(\delta^2)), g_\infty)} \geq 1 - \varepsilon. \qquad \square$$

The convergence in $C^1(M \setminus B_y(\delta_0))$ implies strong convergence in $L^2(M \setminus B_y(\delta_0), g_\infty)$ of $\psi_{i,\ell}$ to $\bar{\psi}_\ell$. Hence

$$\|\bar{\psi}_\ell\|_{L^2(M \setminus B_y(\delta_0), g_\infty)} \geq 1 - \varepsilon,$$

and thus $\|\bar{\psi}_\ell\|_{L^2(M_\infty, g_\infty)} = 1$. The orthogonality of these sections is provided by the following lemma, and the inequality (1.13) then follows immediately.

Lemma 1.4.4 *The sections $\bar{\psi}_1, \ldots, \bar{\psi}_m$ are orthogonal.*

Proof of the lemma. The sections $\varphi_{i,1}, \ldots, \varphi_{i,\ell}$ are orthogonal. For any fixed δ_0 (given by the previous lemma), it follows for sufficiently large i that

$$
\begin{aligned}
\left| \int_{M \setminus B_y(\delta_0)} \langle \psi_{i,\ell}, \psi_{i,\tilde{\ell}} \rangle \, dv^{g_\infty} \right| &= \left| \int_{M \setminus B_y(\delta_0)} \langle \varphi_{i,\ell}, \varphi_{i,\tilde{\ell}} \rangle \, dv^{g_{L_i}} \right| \\
&= \left| \int_{B_y(\delta_0)} \langle \varphi_{i,\ell}, \varphi_{i,\tilde{\ell}} \rangle \, dv^{g_{L_i}} \right| \\
&= \left| \int_{B_y(\delta_0)} \underbrace{\left(\frac{F_{L_i}}{F_\infty} \right)^k}_{\leq 1} \langle \psi_{i,\ell}, \psi_{i,\tilde{\ell}} \rangle \, dv^{g_\infty} \right| \\
&\leq \varepsilon^2
\end{aligned}
\tag{1.16}
$$

Because of strong L^2 convergence on $M \setminus B_y(\delta_0)$ this implies

$$
\left| \int_{M \setminus B_y(\delta_0)} \langle \bar{\psi}_\ell, \bar{\psi}_{\tilde{\ell}} \rangle \, dv^{g_\infty} \right| \leq \varepsilon^2
\tag{1.17}
$$

for $\tilde{\ell} \neq \ell$, and hence in the limit $\varepsilon \to 0$ (and $\delta_0 \to 0$) we get the orthogonality of $\bar{\psi}_1, \ldots, \bar{\psi}_m$. $\square$

Appendix A   Analysis on (M_∞, g_∞)

The aim of this appendix is to sketch how to prove Proposition 1.3.1. All properties in this appendix are well-known to experts, but explicit references are not evident to find. Thus this summary might be helpful to the reader.

The geometry of (M_∞, g_∞) is asymptotically cylindrical. The metric g_∞ is even a b-metric in the sense of Melrose [31], but to keep the presentation simple, we avoid the b-calculus.

If $(r, \gamma) \in \mathbb{R}^+ \times \mathbb{S}^{n-1}$ denote polar normal coordinates with respect to the metric g_0, and if we set $t := -\log r$, then (t, γ) defines a diffeomorphism $\alpha :$ $B_y^{(M, g_0)}(1/2) \setminus \{y\} \to [\log 2, \infty) \times \mathbb{S}^{n-1}$ such that $(\alpha^{-1})^* g_\infty = dt^2 + h_t$ for a family of metrics such that $(\alpha^{-1})^* g_\infty$, all of its derivatives, its curvature, and all derivatives of the curvature tend to the standard metric on the cylinder, and the speed of the convergence is majorised by a multiple of e^t. Thus the continuity of the coefficients property implies, that P_∞ extends to a bounded operator from $\Gamma_{H^k(M_\infty, g_\infty)}(V) \to \Gamma_{L^2(M_\infty, g_\infty)}(V)$.

The formal self-adjointness of P_∞ implies that

$$\int_{M_\infty} \langle \psi, P_\infty \varphi \rangle = \int_{M_\infty} \langle P_\infty \psi, \varphi \rangle \tag{A.18}$$

holds for $\varphi, \psi \in \Gamma_c(V)$ and as $\Gamma_c(V)$ is dense in H^k, property (A.18) follows all H^k-sections φ, ψ.

To show Proposition 1.3.1 it remains to prove the regularity estimate and then to verify that the adjoint of $P_\infty : \Gamma_{H^k(M_\infty, g_\infty)}(V) \to \Gamma_{L^2(M_\infty, g_\infty)}(V)$ has domain $\Gamma_{H^k(M_\infty, g_\infty)}(V)$.

For proving the regularity estimate we need the following local estimate.

Lemma A.1 *Let K be a compact subset of a Riemannian manifold (U, g). Let P be an elliptic differential operator on U of order $k \geq 1$. Then there is a*

19

constant $C = C(U, K, P, g)$ such that

$$\|u\|_{H^k(K,g)} \le C\big(\|u\|_{L^2(U,g)} + \|Pu\|_{L^2(U,g)}\big). \tag{A.19}$$

Here the $H^k(K, g)$-norm is defined via the Levi-Civita connection for g.

This estimate holds uniformly in an ε-neighborhood of P and g in the following sense. Assume that $\tilde{P}$ is another differential operator, and that the C^0-norm of the coefficients of $\tilde{P} - P$ is at most ε, where ε is small. Also assume that $\tilde{g}$ is ε-close to g in the C^k-topology. Then the estimate (A.19) holds for $\tilde{P}$ instead of P and for $\tilde{g}$ instead of g and again for a constant $C = C(U, K, P, g, \varepsilon)$.

Proof of the lemma. We cover the compact set K by a finite number of coordinate neighborhoods $U_1, \ldots, U_m$. We choose open sets $V_i \subset U_i$ such that the closure of V_i is compact in U_i and such that $K \subset V_1 \cup \ldots \cup V_m$. One can choose compact sets $K_i \subset V_i$ such that $K = K_1 \cup \ldots \cup K_m$. To prove (A.19) it is sufficient to prove $\|u\|_{H^k(K_i,g)} \le C(\|u\|_{L^2(V_i,g)} + \|Pu\|_{L^2(V_i,g)})$ for any i.

We write this inequality in coordinates. As the closure of V_i is a compactum in U_i, the transition to coordinates changes the above inequality only by a constant. The operator P, written in a coordinate chart is again elliptic.

We have thus reduced the prove of (A.19) to the prove of the special case that U and K are open subsets of flat $\mathbb{R}^n$.

The proof of this special case is explained in detail for example in in [33, Corollary III 1.5]. The idea is to construct a parametrix for P, i.e. a pseudodifferential operator of order $-k$ such that $S_1 := QP - \mathrm{Id}$ and $S_2 := PQ - \mathrm{Id}$ are infinitely smoothing operators. Thus Q is bounded from $L^2(U)$ to the Sobolev space $H^k(U)$, in particular $\|Q(P(u))\|_{H^k} \le C\|P(u)\|_{L^2}$. Smoothing operators map the Sobolev space L^2 continuously to H^k. We obtain

$$\|u\|_{H^k(K)} \le \|u\|_{H^k(U)} \le \|Q(P(u))\|_{H^k(U)} + \|S_1(u)\|_{H^k(U)}$$
$$\le C'\big(\|P(u)\|_{L^2(U)} + \|u\|_{L^2(U)}\big).$$

See also [28, III §3] for a good presentation on how to construct and work with such a parametrix.

To see the uniformicity, one verifies that

$$\left|\frac{\|u\|_{H^k(K,\tilde{g})}}{\|u\|_{H^k(K,g)}} - 1\right| \le C \|\tilde{g} - g\|_{C^k} \le C\varepsilon$$

and

$$\left|\frac{\|\tilde{P}(u)\|_{L^2(U)}}{\|P(u)\|_{L^2(U)}} - 1\right| \le C\varepsilon \|u\|_{H^k(U)}.$$

The unformicity statement thus follows. $\qquad\square$

Proof of the regularity estimate in Proposition 1.3.1. We write M_∞ as $M_B \cup ([0, \infty) \times S^{n-1})$, such that the metric g_∞ is asymptotic (in the C^∞-sense) to the standard cylindrical metric. The metric g_∞ restricted to $[R - 1, R + 2] \times S^{n-1}$ then converges in the C^k-topology to the cylindrical metric $dt^2 + \sigma^{n-1}$ on $[0, 3] \times S^{n-1}$ for $R \to \infty$. As the coefficients of P_g depend continuously on the metric, the P-operators on $[R - 1, R + 2] \times S^{n-1}$ is in an ε-neighborhood of P, for $R \geq R_0 = R_0(\varepsilon)$. Applying the preceding lemma for $K = [R, R + 1] \times S^{n-1}$ and $U = (R - 1, R + 2) \times S^{n-1}$ we obtain

$$\|\nabla^s u\|_{L^2([R, R+1] \times S^{n-1}, g_\infty)} \leq C \big(\|u\|_{L^2((R-1, R+2) \times S^{n-1}, g_\infty)}$$
$$+ \|P_\infty u\|_{L^2((R-1, R+2) \times S^{n-1}, g_\infty)} \big). \qquad (A.20)$$

Similarly, applying the lemma to $K = M_B \cup ([0, R_0] \times S^{n-1})$ and $U = M_B \cup ([0, R_0 + 1) \times S^{n-1})$ gives

$$\|\nabla^s u\|_{L^2(M_B \cup ([0, R_0] \times S^{n-1}), g_\infty)} \leq C \big(\|u\|_{L^2(M_B \cup ([0, R_0+1) \times S^{n-1}), g_\infty)}$$
$$+ \|P_\infty u\|_{L^2(M_B \cup ([0, R_0+1) \times S^{n-1}), g_\infty)} \big). \qquad (A.21)$$

Taking the sum of estimate (A.21), of estimate (A.20) for $R = R_0$, again estimate (A.20) but for $R = R_0 + 1$, and so for all $R \in \{R_0 + 2, R_0 + 3, \ldots\}$ we obtain (1.9), with a larger constant C. $\qquad \square$

Now we study the domain $\mathcal{D}$ of the adjoint of

$$P_\infty : \Gamma_{H^k(M_\infty, g_\infty)}(V) \to \Gamma_{L^2(M_\infty, g_\infty)}(V).$$

By definition a section $\varphi : \Gamma_{L^2(M_\infty, g_\infty)}(V)$ is in $\mathcal{D}$ if and only if

$$\Gamma_{H^k(M_\infty, g_\infty)}(V) \ni u \mapsto \int_{M_\infty} \langle P_\infty u, \varphi \rangle \qquad (A.22)$$

is bounded as a map from L^2 to $\mathbb{R}$. For $\varphi \in \Gamma_{H^k(M_\infty, g_\infty)}(V)$ we know that $P_\infty \varphi$ is L^2 and thus property (A.18) directly implies this boundedness. Thus $\Gamma_{H^k(M_\infty, g_\infty)}(V) \subset \mathcal{D}$.

Conversely assume the boundedness of (A.22). Then there is a $v \in \Gamma_{L^2(M_\infty, g_\infty)}(V)$ such that $\int_{M_\infty} \langle u, v \rangle = \int_{M_\infty} \langle P_\infty u, \varphi \rangle$, or in other words $P_\infty \varphi = v$ holds weakly. Standard regularity theory implies

$$\varphi \in \Gamma_{H^k(M_\infty, g_\infty)}(V).$$

We obtain $\Gamma_{H^k(M_\infty, g_\infty)}(V) = \mathcal{D}$, and thus the self-adjointness of P_∞ follows. Proposition 1.3.1 is thus shown.

References

[1] B. Ammann, *The Dirac Operator on Collapsing Circle Bundles*, Sém. Th. Spec. Géom Inst. Fourier Grenoble **16** (1998), 33–42.

[2] B. Ammann, *Spin-Strukturen und das Spektrum des Dirac-Operators*, Ph.D. thesis, University of Freiburg, Germany, 1998, Shaker-Verlag Aachen 1998, ISBN 3-8265-4282-7.

[3] _____, *The smallest Dirac eigenvalue in a spin-conformal class and cmc-immersions*, Comm. Anal. Geom. **17** (2009), 429–479.

[4] _____, *A spin-conformal lower bound of the first positive Dirac eigenvalue*, Diff. Geom. Appl. **18** (2003), 21–32.

[5] _____, *A variational problem in conformal spin geometry*, Habilitationsschrift, Universität Hamburg, 2003.

[6] B. Ammann and C. Bär, *Dirac eigenvalues and total scalar curvature*, J. Geom. Phys. **33** (2000), 229–234.

[7] B. Ammann and E. Humbert, *The first conformal Dirac eigenvalue on 2-dimensional tori*, J. Geom. Phys. **56** (2006), 623–642.

[8] B. Ammann, E. Humbert, and B. Morel, *Mass endomorphism and spinorial Yamabe type problems*, Comm. Anal. Geom. **14** (2006), 163–182.

[9] B. Ammann, A. D. Ionescu, and V. Nistor, *Sobolev spaces on Lie manifolds and regularity for polyhedral domains*, Doc. Math. **11** (2006), 161–206.

[10] B. Ammann, R. Lauter, and V. Nistor, *On the geometry of Riemannian manifolds with a Lie structure at infinity*, Int. J. Math. Math. Sci. (2004), 161–193.

[11] _____, *Pseudodifferential operators on manifolds with a Lie structure at infinity*, Ann. of Math. **165** (2007), 717–747.

[12] C. Bär, *The Dirac operator on space forms of positive curvature*, J. Math. Soc. Japan **48** (1996), 69–83.

[13] C. Bär, *The Dirac operator on hyperbolic manifolds of finite volume*, J. Differ. Geom. **54** (2000), 439–488.

[14] H. Baum, *Spin-Strukturen und Dirac-Operatoren über pseudoriemannschen Mannigfaltigkeiten*, Teubner Verlag, 1981.

[15] A. L. Besse, *Einstein manifolds*, Ergebnisse der Mathematik und ihrer Grenzgebiete, 3. Folge, no. 10, Springer-Verlag, 1987.

[16] T. P. Branson, *Differential operators canonically associated to a conformal structure*, Math. Scand. **57** (1985), 293–345.

[17] _____, *Group representations arising from Lorentz conformal geometry*, J. Funct. Anal. **74** (1987), no. 2, 199–291.

[18] _____, *Second order conformal covariants*, Proc. Amer. Math. Soc. **126** (1998), 1031–1042.

[19] B. Colbois and A. El Soufi, *Eigenvalues of the Laplacian acting on p-forms and metric conformal deformations*, Proc. of Am. Math. Soc. **134** (2006), 715–721.

[20] A. El Soufi and S. Ilias, *Immersions minimales, première valeur propre du laplacien et volume conforme*, Math. Ann. **275** (1986), 257–267.

[21] H. D. Fegan, *Conformally invariant first order differential operators*, Quart. J. Math. Oxford, II. series **27** (1976), 371–378.

[22] R. Gover and L. J. Peterson, *Conformally invariant powers of the Laplacian, Q-curvature, and tractor calculus*, Comm. Math. Phys. **235** (2003), 339–378.

[23] E. Hebey and F. Robert, *Coercivity and Struwe's compactness for Paneitz type operators with constant coefficients*, Calc. Var. Partial Differential Equations **13** (2001), 491–517.

[24] N. Hitchin, *Harmonic spinors*, Adv. Math. **14** (1974), 1–55.

[25] P. Jammes, *Extrema de valeurs propres dans une classe conforme*, Sémin. Théor. Spectr. Géom. **24** (2007), 23–42.

[26] I. Kolář, P. W. Michor, and J. Slovák, *Natural operations in differential geometry*, Springer-Verlag, Berlin, 1993.

[27] N. Korevaar, *Upper bounds for eigenvalues of conformal metrics*, J. Differ. Geom. **37** (1993), 73–93.

[28] H.-B. Lawson and M.-L. Michelsohn, *Spin Geometry*, Princeton University Press, 1989.

[29] J. M. Lee and T. H. Parker. *The Yamabe problem.* Bull. Am. Math. Soc., New Ser. **17** (1987), 37–91.

[30] S. Mac Lane, *Categories for the working mathematician*, Graduate Texts in Mathematics, vol. 5, Springer-Verlag, New York, 1998.

[31] R. B. Melrose, *The Atiyah-Patodi-Singer index theorem*, Research Notes in Mathematics, vol. 4, A K Peters Ltd., Wellesley, MA, 1993.

[32] S. M. Paneitz, *A quartic conformally covariant differential operator for arbitrary pseudo-Riemannian manifolds*, Preprint 1983, published in SIGMA **4** (2008).

[33] M. E. Taylor, M. E., *Pseudodifferential operators*, Princeton University Press, 1981.

Authors' addresses:

Bernd Ammann
Facultät für Mathematik
Universität Regensburg
93040 Regensburg
Germany
bernd.ammann@mathematik.uni-regensburg.de

Pierre Jammes
Laboratoire J.-A. Diendonné, Université Nice-Sophia Antipolis,
Parc Valrose, F-06108 Nice Cedex02, France
pjammes@unice.fr

2

K-Destabilizing test configurations with smooth central fiber

CLAUDIO AREZZO, ALBERTO DELLA VEDOVA, AND
GABRIELE LA NAVE

Abstract

In this note we point out a simple application of a result by the authors in
[2]. We show how to construct many families of strictly K-unstable polarized
manifolds, destabilized by test configurations with smooth central fiber. The
effect of resolving singularities of the central fiber of a given test configuration
is studied, providing many new examples of manifolds which do not admit
Kähler constant scalar curvature metrics in some classes.

2.1 Introduction

In this note we want to speculate about the following Conjecture due to Tian-
Yau-Donaldson ([23], [24], [25], [7]):

Conjecture 2.1.1 *A polarized manifold* (M, A) *admits a Kähler metric of
constant scalar curvature in the class* $c_1(A)$ *if and only if it is* K-*polystable.*

The notion of K-stability will be recalled below. For the moment it suffices to
say, loosely speaking, that a polarized manifold, or more generally a polarized
variety (V, A), is K-stable if and only if any *special* degeneration or *test
configuration* of (V, A) has an associated *non positive* weight, called Futaki
invariant and that this is zero only for the product configuration, i.e. the trivial
degeneration.

We do not even attempt to give a survey of results about Conjecture 2.1.1, but
as far as the results of this note are concerned, it is important to recall the reader
that Tian [24], Donaldson [7], Stoppa [22], using the results in [3] and [4], and
Mabuchi [17] have proved the sufficiency part of the Conjecture. Destabilizing
a polarizing manifold then implies non existence results of Kähler constant
scalar curvature metrics in the corresponding classes.

One of the main problems in this subject is that under a special degeneration a smooth manifold often becomes very singular, in fact just a polarized scheme in general. This makes all the analytic tool available at present very difficult to use.

Hence one naturally asks which type of singularities must be introduced to make the least effort to destabilize a smooth manifold without cscK metrics.

The aim of this note is to provide a large class of examples of special degenerations with positive Futaki invariant and *smooth* limit. In fact we want to provide a "machine" which associates to any special degeneration of a polarized *normal* variety (V, A) with positive Futaki invariant a special degeneration *for a polarized manifold* $(\tilde{M}, \tilde{A})$ with smooth central fiber and still positive Futaki invariant.

To the best of our knowledge, before this work the only known examples of special degeneration with non negative Futaki invariant and smooth central fiber are the celebrated example of Mukai-Umemura's Fano threefold ([18]) used by Tian in [24] to exhibit the first examples of Fano manifolds with discrete automorphism group and no Kähler-Einstein metrics (other Fano manifolds with these properties have been then produced in [1]). In this case there exist non trivial special degenerations with smooth limit and *zero* Futaki invariant (hence violating the definition of K-stability). It then falls in the borderline case, making this example extremely interesting and delicate. We stress that our "machine" does not work in this borderline case, because a priori the Futaki invariant of the new test configuration is certainly small (by [2]) but we cannot control its sign.

To state our result more precisely we now recall the relevant definitions:

Definition 2.1.2 Let (V, A) be a n-dimensional polarized variety or scheme. Given a one-parameter subgroup $\rho : \mathbb{C}^* \to \mathrm{Aut}(V)$ with a linearization on A and denoted by $w(V, A)$ the weight of the $\mathbb{C}^*$-action induced on $\bigwedge^{\mathrm{top}} H^0(V, A)$, we have the following asymptotic expansions as $k \gg 0$:

$$h^0(V, A^k) = a_0 k^n + a_1 k^{n-1} + O(k^{n-2}) \tag{2.1}$$

$$w(V, A^k) = b_0 k^{n+1} + b_1 k^n + O(k^{n-1}) \tag{2.2}$$

The (normalized) *Futaki invariant* of the action is the rational number

$$F(V, A, \rho) = \frac{b_1}{a_0} - \frac{b_0 a_1}{a_0^2}.$$

Definition 2.1.3 A *test configuration* (X, L) for a polarized variety (V, A) consists of a scheme X endowed with a $\mathbb{C}^*$-action that linearizes on a line bundle L over X, and a flat $\mathbb{C}^*$-equivariant map $f : X \to \mathbb{C}$ (where $\mathbb{C}$ has the usual weight one $\mathbb{C}^*$-action) such that $L|_{f^{-1}(0)}$ is ample on $f^{-1}(0)$ and we have $(f^{-1}(1), L|_{f^{-1}(1)}) \simeq (V, A^r)$ for some $r > 0$.

When (V, A) has a $\mathbb{C}^*$-action $\rho \colon \mathbb{C}^* \to \text{Aut}(V)$, a test configuration where $X = V \times \mathbb{C}$ and $\mathbb{C}^*$ acts on X diagonally through ρ is called *product configuration*.

Given a test configuration (X, L) we will denote by $F(X, L)$ the Futaki invariant of the $\mathbb{C}^*$-action induced on the central fiber $(f^{-1}(0), L|_{f^{-1}(0)})$. If (X, L) is a product configuration as above, clearly we have $F(X, L) = F(V, A, \rho)$.

Definition 2.1.4 The polarized manifold (M, A) is *K-stable* if for each test configuration for (M, A) the Futaki invariant of the induced action on the central fiber $(f^{-1}(0), L|_{f^{-1}(0)})$ is less than or equal to zero, with equality if and only if we have a product configuration.

A test configuration (X, L) is called *destabilizing* if the Futaki invariant of the induced action on $(f^{-1}(0), L|_{f^{-1}(0)})$ is greater than zero.

Test configurations for an embedded variety $V \subset \mathbb{P}^N$ endowed with the hyperplane polarization A can be constructed as follows. Given a one-parameter subgroup $\rho \colon \mathbb{C}^* \to GL(N + 1)$, which induces an obvious diagonal $\mathbb{C}^*$-action on $\mathbb{P}^N \times \mathbb{C}$, it clear that the subscheme

$$X = \overline{\left\{ (z, t) \in \mathbb{P}^N \times \mathbb{C} \,|\, t \neq 0, \, (\rho(t^{-1})z, t) \in V \right\}} \subset \mathbb{P}^N \times \mathbb{C},$$

is invariant and projects equivariantly on $\mathbb{C}$. Thus considering the relatively ample polarization L induced by the hyperplane bundle gives test configuration for (V, A). On the other hand, given a test configuration (X, L) for a polarized variety (V, A), the relative projective embedding given by L^r, with r sufficiently large, realizes X as above (see details in [21]).

We can now describe our "machine": consider a test configuration (X, L) for a polarized normal variety (V, A) with $F(X, L) > 0$. Up to raise L to a suitable power – which does not affect the Futaki invariant – we can suppose being in the situation above with $X \subset \mathbb{P}^N \times \mathbb{C}$ invariantly, and L induced by the hyperplane bundle of $\mathbb{P}^N$. At this point we consider the central fiber $X_0 \subset \mathbb{P}^N$, which is invariant with respect to ρ, and we apply the (equivariant) resolution of singularities [14, Corollary 3.22 and Proposition 3.9.1]. Thus there is a smooth manifold $\tilde{P}$ acted on by $\mathbb{C}^*$ and an equivariant map

$$\beta \colon \tilde{P} \to \mathbb{P}^N$$

which factorizes through a sequence of blow-ups, such that the strict transform $\tilde{X}_0$ of X_0 is invariant and smooth. The key observation is that the strict transform $\tilde{X}_1$ of the fiber $X_1 \subset X$ degenerate to X_0 under the given $\mathbb{C}^*$ action on $\tilde{P}$, thus it must be smooth. This gives an invariant family $\tilde{X} \subset P \times \mathbb{C}$ and an equivariant birational morphism

$$\pi : \tilde{X} \to X.$$

Some comments are in order:

1 *all* the fibers of $\tilde{X}$ are smooth, but π is never a resolution of singularities of X (except the trivial case when the central fiber of X was already smooth) since it fails to be an isomorphism on the smooth locus of X;
2 $\tilde{L} = \pi^* L$ is not a relatively ample line bundle any more, but just a big and nef one. It is not then even clear what it means to compute its Futaki invariant;
3 the fiber over the generic point of $\mathbb{C}$ of the new (big and nef) test configuration $(\tilde{X}, \tilde{L})$ is different from V;
4 the family $\tilde{X}$ is not unique since the resolution β it is not.

The issue raised at point (2) was addressed in [2] and it was proved that the following natural (topological) definition makes the Futaki invariant a continuous function around big and nef points in the Kähler cone. We will give simple self-contained proofs in the cases of smooth manifolds and varieties with just normal singularities in Section 2.

Definition 2.1.5 Let V be a projective variety or scheme endowed with a $\mathbb{C}^*$-action and let B be a big and nef line bundle on V. Choosing a linearization of the action on B gives a $\mathbb{C}^*$-representation on $\bigoplus_{j=0}^{\dim V} H^j(V, B^k)^{(-1)^j}$ (here the E^{-1} denotes the dual of E). We set $w(V, B^k) = \operatorname{tr} A_k$, where A_k is the generator of that representation. As $k \to +\infty$ we have the following expansion

$$\frac{w(V, B^k)}{\chi(V, B^k)} = F_0 k + F_1 + O(k^{-1}),$$

and we define

$$F(V, B) = F_1$$

to be the *Donaldson–Futaki invariant* of the chosen action on (V, B)

The existence of the expansion involved in definition above follows from the standard fact that $\chi(V, B^k)$ is a polynomial of degree $\dim V$, whose proof (see for example [11]) can be easily adapted to show that $w(V, B^k)$ is a polynomial of degree at most $\dim V + 1$.

The key technical Theorem proved in [2] is then the following:

Theorem 2.1.6 *Let B, A be linearized line bunldes on a scheme V acted on by $\mathbb{C}^*$. Suppose that B is big and nef and A ample. We have*

$$F(V, B^r \otimes A) = F(V, B) + O\left(\frac{1}{r}\right), \qquad as\ r \to \infty.$$

Having established a good continuity property of the Futaki invariant up to these boundary point, we need to address the question of the effect of a resolution of singularities of the central fiber. This is a particular case of the following non trivial extension of previous analysis by Ross and Thomas [21] which was proved in [2] where the general case of birational morphisms has been studied:

Theorem 2.1.7 *Given a test configuration $f : (X, L) \to \mathbb{C}$ as above, let $f' : (X', L') \to \mathbb{C}$ be another flat equivariant family with X' normal and let $\beta : (X', L') \to (X, L)$ be a $\mathbb{C}^*$-equivariant birational morphism such that $f' = f \circ \beta$ and $L' = \beta^* L$. Then we have*

$$F(X', L') \geq F(X, L),$$

with strict inequality if and only if the support of $\beta_*(\mathcal{O}_{X'})/\mathcal{O}_X$ has codimension one.

The proof of these results uses some heavy algebraic machinery, yet their proof when (V, A) or the central fiber of (X, L) have only normal singularities (a case largely studied) is quite simple and we give it in Section 2.

The Corollary of Theorem 2.1.6 and Theorem 2.1.7 we want to point out in this note is then the following:

Theorem 2.1.8 *Let (X, L) be a test configuration for the polarized normal variety (V, A) with positive Futaki invariant. Let moreover $(\tilde{X}, \tilde{L})$ be a (big and nef) test configuration obtained from (X, L) as above and let $(\tilde{M}, \tilde{B})$ be the smooth (big and nef) fiber over the point $1 \in \mathbb{C}$. Let R be any relatively ample line bundle over $\tilde{X}$.*

Then $(\tilde{X}, \tilde{L}^r \otimes R)$ is a test configuration for $(\tilde{M}, \tilde{B}^r \otimes R|_{\tilde{M}})$ with following properties:

1 smooth central fiber;
2 positive Futaki invariant for r sufficiently large.

In particular $\tilde{M}$ does not admit a constant scalar curvature Kähler metric in any class of the form $c_1(\tilde{B}^r \otimes R|_{\tilde{M}})$, with r large enough.

While this Theorem clearly follows from Theorems 2.1.6 and Theorem 2.1.7, but for the specific case of central fiber with normal singularities it follows from the much simpler Proposition 2.2.1 and Theorem 2.2.3.

The range of applicability of the above theorem is very large. We go through the steps of the resolution of singularities in an explicit example by Ding-Tian [6] of a complex orbifold of dimension 2. In this simple example explicit calculations are easy to perform, yet we point out that the final example is somehow trivial since it ends on a product test configuration. On the other hand abundance of similar examples even in dimension 2 can be obtained by the reader as an exercise using the results of Jeffres [12] and Nagakawa [19], in which cases we loose an explicit description of the resulting destabilized manifold, but we get new nontrivial examples. In fact in higher dimensions one can use the approach described in this note to test also the Arezzo-Pacard blow up theorems [3] [4], when the resolution of singularities requires a blow up of a scheme of positive dimension.

2.2 The case of normal singularities

In this section we give simple proofs of the continuity of the Futaki invariant at boundary points for smooth manifolds or varieties with normal singularities. More general results of this type have been proved in [2] but we want to stress that under these assumptions proofs become much easier.

The fundamental continuity property we will need, and proved in Corollary 2.1.6, can be stated in the following form for smooth bases:

Proposition 2.2.1 *Let A, L be respectively an ample and a big and nef line bundle on a smooth projective manifold M. For every $\mathbb{C}^*$-action on M that linearizes to A and L, as $r \to +\infty$ we have*

$$F(M, L^r \otimes A) = F(M, L) + O\left(\frac{1}{r}\right).$$

Proof The result is a simple application of the equivariant Riemann-Roch Theorem. We present here the details of the calculations involved, since we could not find precise references for them.

Fix an hermitian metrics on A that is invariant with respect to the action of $S^1 \subset \mathbb{C}^*$ and suppose that the curvature ω is a Kähler metric. Since L is nef, for each $r > 0$ we can choose an invariant metric on L whose curvature η_r satisfy $r\eta_r + \omega > 0$. In other words $r\eta_r + \omega$ is a Kähler form which coincides with the curvature of the induced hermitian metric on the line bundle $L^r \otimes A$.

Setting $n = \dim(M)$, by Riemann–Roch for each $r > 0$ we have:

$$\chi(M, L^{rk} \otimes A^k) = (rk)^n \int_M \frac{(\eta_r + \frac{1}{r}\omega)^n}{n!}$$

$$+ (rk)^{n-1} \int_M \frac{(\eta_r + \frac{1}{r}\omega)^{n-1} \wedge \mathrm{Ric}(\eta_r + \frac{1}{r}\omega)}{2(n-1)!} + O(k^{n-2}).$$

Now let v be the holomorphic vector field on M generating the given $\mathbb{C}^*$-action. Let f and g_r be smooth S^1-invariant functions on M such that $i_v\omega + \bar{\partial}f = 0$ and $i_v\eta_r + \bar{\partial}g_r = 0$. They exist since the given $\mathbb{C}^*$-action on M lifts to A and $L^r \otimes A$ (see [7, pag. 294]). Applying, in a similar fashion, the equivariant Riemann-Roch theorem, the weight of the induced action on $\bigotimes_{j=0}^n \det H^j(M, L^{rk} \otimes A^k)^{(-1)^j}$ is calculated as:

$$w(M, L^{rk} \otimes A^k) = (rk)^{n+1} \int_M \left(g_r + \frac{1}{r}f\right) \frac{(\eta_r + \frac{1}{r}\omega)^n}{n!}$$

$$+ (rk)^n \int_M \left(g_r + \frac{1}{r}f\right) \frac{(\eta_r + \frac{1}{r}\omega)^{n-1} \wedge \mathrm{Ric}(\eta_r + \frac{1}{r}\omega)}{2(n-1)!}$$

$$+ O(k^{n-1}).$$

As $r \to +\infty$ we have expansions:

$$\int_M \frac{(\eta_r + \frac{1}{r}\omega)^n}{n!} = \frac{c_1(L^r \otimes A)^n}{n!\, r^n} = \frac{c_1(L)^n}{n!} + O\left(\frac{1}{r}\right)$$

$$\int_M \frac{(\eta_r + \frac{1}{r}\omega)^{n-1} \wedge \mathrm{Ric}(\eta_r + \frac{1}{r}\omega)}{2(n-1)!} = \frac{c_1(L^r \otimes A)^{n-1}c_1(M)}{2(n-1)!\, r^{n-1}} + O\left(\frac{1}{r}\right)$$

$$= \frac{c_1(L)^{n-1}c_1(M)}{2(n-1)!} + O\left(\frac{1}{r}\right)$$

$$\int_M \left(g_r + \frac{1}{r}f\right) \frac{(\eta_r + \frac{1}{r}\omega)^n}{n!} = \frac{c_1^T(L^r \otimes A)^{n+1}}{(n+1)!\, r^{n+1}}$$

$$= \frac{c_1^T(L)^{n+1}}{(n+1)!} + O\left(\frac{1}{r}\right)$$

$$\int_M \left(g_r + \frac{1}{r}f\right) \frac{(\eta_r + \frac{1}{r}\omega)^{n-1} \wedge \mathrm{Ric}(\eta_r + \frac{1}{r}\omega)}{2(n-1)!} = \frac{c_1^T(L^r \otimes A)^n c_1^T(M)}{2n!\, r^n}$$

$$= \frac{c_1^T(L)^n c_1^T(M)}{2n!} + O\left(\frac{1}{r}\right),$$

where c_1^T denotes the equivariant first Chern class. Thus we have:

$$F(M, L^r \otimes A) = \frac{\frac{c_1^T(L)^n c_1^T(M)}{2n!} \cdot \frac{c_1(L)^n}{n!} - \frac{c_1^T(L)^{n+1}}{(n+1)!} \cdot \frac{c_1(L)^{n-1}c_1(M)}{2(n-1)!}}{\left(\frac{c_1(L)^n}{n!}\right)^2} + O\left(\frac{1}{r}\right).$$

and the thesis follows applying again the (equivariant) Riemann-Roch theorem to L. $\square$

The following has essentially been proved by Paul-Tian ([20]); here we give a simple self-contained proof.

Proposition 2.2.2 *Let (V, L) be a polarized normal projective variety endowed with a $\mathbb{C}^*$-action. Let $\pi : \tilde{V} \to V$ be an equivariant resolution of singularities. We have*

$$F(V, L) = F(\tilde{V}, \pi^* L).$$

Proof By Zariski's Main theorem we have an equivariant isomorphism $\pi_* \mathcal{O}_{\tilde{V}} \simeq \mathcal{O}_V$, moreover normality of V implies $\dim \operatorname{Supp} R^q \pi_*(\mathcal{O}_{\tilde{V}}) \le n - 1$ for all $q > 0$, where $n + 1 = \dim V = \dim \tilde{V}$. In fact, since π is an isomorphisms outside the singularities of V, $\operatorname{Supp} R^q \pi_*(\mathcal{O}_{\tilde{V}})$ is contained in the singular locus of V, which by normality has dimension less than or equal to $n - 1$.

Thus the projection formula yields:

$$\chi(\tilde{V}, \pi^* L^k) = \chi(V, R\pi_*(\mathcal{O}_{\tilde{V}}) \otimes L^k) - \chi(V, L^k) + O(k^{n-2}),$$

and by the same token:

$$w(\tilde{V}, \pi^* L^k) = w(V, L^k) + O(k^{n-1}).$$

The thesis follows by definition 2.1.5. $\square$

Combining Propositions 2.2.1 and 2.2.2 we get the following result, which explains the behaviour for normal varieties of the Futaki invariant under Hironaka's resolution of singularities process:

Theorem 2.2.3 *Let (V, L) be a polarized normal variety endowed with a $\mathbb{C}^*$-action. Let $\pi : \tilde{V} \to V$ be an equivariant resolution of singularities with exceptional divisor E. Then we have*

$$F\left(\tilde{V}, \pi^* L^r \otimes \mathcal{O}(-E)\right) = F(V, L) + O\left(\frac{1}{r}\right) \qquad as \ r \to +\infty.$$

In particular, if $F(V, L) \neq 0$, then $\tilde{V}$ admits no cscK metrics in the classes $r\pi^*c_1(L) - E$ for $r \gg 0$.

Proof Since π is a sequence of blow-ups along smooth centers, π^*L is big and nef, moreover there exists $r_0 > 0$ such that $\pi^*L^{r_0} \otimes \mathcal{O}(-E)$ is ample. Thus, as $r \to +\infty$ we obtain:

$$F\left(\tilde{V}, \pi^*L^r \otimes \mathcal{O}(-E)\right) = F\left(\tilde{V}, \pi^*L^{r-r_0} \otimes \pi^*L^{r_0} \otimes \mathcal{O}(-E)\right)$$

$$= F\left(\tilde{V}, \pi^*L^{r-r_0}\right) + O\left(\frac{1}{r}\right)$$

$$= F(V, L) + O\left(\frac{1}{r}\right),$$

where the second and third equalities follow from Propositions 2.2.1 and 2.2.2 respectively. □

2.3 Proof of Theorem 2.1.8 and examples

Proof of Theorem 2.1.8 On each fiber of $\tilde{X} \to \mathbb{C}$ the restrictions of $\tilde{L}^r$ and R are nef and ample line bundles respectively, thus $\tilde{L}^r \otimes R$ is relatively ample and gives, together with X, a test configuration for $(\tilde{M}, \tilde{B}^r \otimes R)$, where $\tilde{B} = \tilde{L}|_{\tilde{M}}$.

All the fibers of that test configuration are smooth by construction thus point (1) is proved. Moreover by Theorem 2.1.6 (or Proposition 2.2.1 since the central fiber of $\tilde{X}$ is smooth) we have the asymptotic expansion as $r \to +\infty$

$$F(\tilde{X}, \tilde{L}^r \otimes R) = F(\tilde{X}, \tilde{L}) + O\left(\frac{1}{r}\right). \tag{2.3}$$

On the other hand, the morphism $\pi : \tilde{X} \to X$ is equivariant and birational, and $\tilde{L} = \pi^*L$, so by Theorem 2.1.7 we get

$$F(\tilde{X}, \tilde{L}) \geq F(X, L) > 0. \tag{2.4}$$

Thus point (2) follows readily from (2.3) and (2.4). □

Let us now see an explicit appearance of the phenomenon described in Theorem 2.1.8. We construct K-destabilizing degenerations with smooth central fiber of a smooth surface obtained by resolution of singularities of a mild singular unstable cubic surface found by Ding-Tian ([6]).

Let us the look at $X_f \subset \mathbb{P}^3$ be the zero locus of $f = z_0z_1^2 + z_2z_3(z_2 - z_3)$.

Let us collect some elementary facts that the reader can easily verify:

1 X_f is singular only at $p_0 = (1:0:0:0)$.
2 Having set (x, y, z) affine co-ordinates centered on $(1:0:0:0)$, let

$$C' = \{((x, y, z), (l_0 : l_1 : l_2)) \in \mathbb{C}^3 \times \mathbb{P}^2 \mid$$
$$xl_1 - yl_0 = xl_2 - zl_0 = yl_2 - zl_1 = 0\}$$

be the blow-up of $\mathbb{C}^3$ at the origin with exceptional divisor $E \simeq \mathbb{P}^2$ and X'_f the proper transform of X_f, then X'_f is singular at points

$$(0:1:0), (0:0:1), (0:1:1) \in E.$$

3 Having set $\tilde{X}_f$ the proper transform of X'_f under the blow-up $\tilde{C}$ of C' at points $(0:1:0), (0:0:1), (0:1:1) \in E$ and E_1 the exceptional divisor over $(0:1:0)$, then $\tilde{X}_f$ is smooth around E_1. Analogously, if E_2 is the exceptional divisor over $(0:0:1)$ and E_3 the one over $(0:1:1)$, then $\tilde{X}_f$ is smooth around E_2 and E_3 too. For future reference let E_0 be the proper transform of E under the second blow up.
4 $\tilde{X}_f$ is smooth.
5 Consider now the $\mathbb{C}^*$-action on $\mathbb{P}^3$ defined by

$$t \cdot (z_0, z_1, z_2, z_3) = (t^{\alpha_0} z_0, t^{\alpha_1} z_1, t^{\alpha_2} z_2, t^{\alpha_3} z_3)$$

with $\alpha_j \in \mathbb{Z}$ and $\alpha_0 + \cdots + \alpha_3 = 0$, thus

$$(t \cdot f)(z) = t^{-(\alpha_0 + 2\alpha_1)} z_0 z_1^2 + t^{-(2\alpha_2 + \alpha_3)} z_2^2 z_3 - t^{-(\alpha_2 + 2\alpha_3)} z_2 z_3^2,$$

and f is semi-invariant if and only if

$$\alpha_0 + 2\alpha_1 = 2\alpha_2 + \alpha_3 = \alpha_2 + 2\alpha_3,$$

hence

$$(\alpha_0, \ldots, \alpha_3) = (-7\beta, 5\beta, \beta, \beta),$$

with $\beta \in \mathbb{Z}$, and f has weight -3β.
6 The monomials of degree three in the variables $z_0, \ldots, z_3$ are a basis of semi-invariants for the fixed $\mathbb{C}^*$-action on $\mathbb{C}[z_0, \ldots, z_3]_3$. In particular fix $\beta = -1$, then the subspace spanned by monomials with weight greater or equal to 4 is

$$V = \text{span}\{z_1^3, z_1^2 z_2, z_1^2 z_3, z_1 z_2^2, z_1 z_2 z_3, z_1 z_3^2\}$$

Thus the hypersurfaces that degenerate to X_f under the fixed $\mathbb{C}^*$-action (with $\beta = -1$) are of the form

$$X_g = \{f + g = 0\},$$

where $g \in V$.

7 The general X_g has a rational double point at $p_0 = (1 : 0 : 0 : 0)$ and is smooth elsewhere.

8 $\tilde{X}_f$ is the flat limit of $\tilde{X}_g$ in $\tilde{P}$, the iterated blow-up of $\mathbb{P}^3$ as described (locally) above.

Thus $\tilde{X}_f$ is the central fiber of the special degeneration of $\tilde{X}_g$ given by the chosen $\mathbb{C}^*$-action. Denoted by $\pi : \tilde{P} \to \mathbb{P}^3$ the blow-up map, let A_r be the restriction of $\pi^* \mathcal{O}_{\mathbb{P}^3}(r) \otimes \mathcal{O}\big(- \sum_{\ell=0}^{4} E_j \big)$ to $\tilde{X}_f$. By Theorem 2.2.3, $F(\tilde{X}_f, A_r)$ has the same sign of $F(X_f, \mathcal{O}_{X_f}(1))$ when r is large enough. But thanks to [6] (see also [16]), with our sign convention, $(X_f, \mathcal{O}_{X_f}(1))$ has positive Futaki invariant, then $\tilde{X}_g$ polarized with the restriction of $\pi^* \mathcal{O}_{\mathbb{P}^3}(r) \otimes \mathcal{O}\big(- \sum_{\ell=0}^{4} E_j \big)$ is K-unstable.

References

[1] C. Arezzo, A. Della Vedova, *On the K-stability of complete intersections in polarized manifolds*, arXiv:0810.1473.

[2] C. Arezzo, A. Della Vedova and G. La Nave, *Singularities and K-semistability*, arXiv:0906.2475.

[3] C. Arezzo and F. Pacard, *Blowing up and desingularizing Kähler orbifolds with constant scalar curvature*, Acta Mathematica **196**, no 2, (2006) 179-228.

[4] C. Arezzo and F. Pacard, *Blowing up Kähler manifolds with constant scalar curvature II*, Annals of Math. **170** no. 2, (2009) 685-738

[5] D. Eisenbud, J. Harris, *The Geometry of schemes*. GTM 197. Springer, 2000.

[6] W. Y. Ding and G. Tian, *Kähler-Einstein metrics and the generalized Futaki invariant*. Invent. Math. **110** (1992), no. 2, 315–335.

[7] S. K. Donaldson, *Scalar curvature and stability of toric varieties*. J. Differential Geom. **62** (2002), no.2, 289–349.

[8] S. K. Donaldson. *Lower bounds on the Calabi functional*, J. Differential Geom., **70** (2005), no.3, 453–472.

[9] J. Fine and J. Ross, *A note on positivity of the CM line bundle*. Int. Math. Res. Not., 2006.

[10] A. Futaki and T. Mabuchi, *Bilinear forms and extremal Kähler vector fields associated with Kähler classes*. Math. Ann. 301 (1995), n.2, 199–210

[11] R. Hartshorne, *Algebraic Geometry*, Springer, 1977.

[12] T. Jeffres, *Singular set of some Kähler orbifolds*. Trans. Amer. Math. Soc. 349 (1997), no. 5, 1961–1971.

[13] G. Kempf, F. Knudsen, D. Mumford and B. Saint-Donat, *Toroidal Embeddings I* Lecture Notes in Mathematics, 339. Springer, 1973.

[14] J. Kollár, *Lectures on resolution of singularities*. Annals of Mathematics Studies, 166. Princeton University Press, Princeton, NJ, 2007.

[15] R. Lazarsfeld. *Positivity in Algebraic Geometry I*. Springer, 2004.

[16] Z. Lu, *On the Futaki invariants of complete intersections*. Duke Math. J. **100** (1999), no.2, 359–372.

[17] T. Mabuchi, *K-stability of constant scalar curvature polarization*, arXiv:0812.4093.

[18] S. Mukai and H. Umemura, *Minimal rational threefolds*. Algebraic geometry (Tokyo/Kyoto, 1982), 490–518, Lecture Notes in Math., 1016, Springer, Berlin, 1983.

[19] Y. Nakagawa, *Bando-Calabi-Futaki characters of Kähler orbifolds*. Math. Ann. 314 (1999), no. 2, 369–380.

[20] S. Paul and G. Tian, *CM Stability and the Generalized Futaki Invariant I*. math.DG/0605278

[21] J. Ross and R. P. Thomas, *An obstruction to the existence of constant scalar curvature Kähler metrics*. Jour. Diff. Geom. **72**, 429–466, 2006.

[22] J. Stoppa, *J. Stoppa, K-stability of constant scalar curvature Kaehler manifolds*, Advances in Mathematics **221** no. 4 (2009), 1397–1408.

[23] G. Tian, *Recent progress on Kähler-Einstein metrics*, in Geometry and physics (Aarhus, 1995), 149-155, Lecture Notes in Pure and Appl. Math., 184, Dekker, New York, 1997.

[24] G. Tian, *Kähler-Einstein metrics with positive scalar curvature*. Invent. Math. **130** (1997), no. 1, 1–37.

[25] G. Tian, *Extremal metrics and geometric stability*. Houston Math. J. **28** (2002), no. 1, 411–432.

Authors' addresses:

Claudio Arezzo
Abdus Salam International Center for Theoretical Physics,
Strada Costiera 11,
Trieste (Italy) and Dipartimento di Matematica,
Università di Parma,
Parco Area delle Scienze 53/A,
Parma, Italy
arezzo@ictp.it

Alberto Della Vedova
Fine Hall,
Princeton University,
Princeton, NJ 08544
and
Dipartimento di Matematica,
Università di Parma,
Parco Area delle Scienze 53/A,
Parma, Italy
della@math.princeton.edu

Gabriele La Nave
Department of Mathematics
Yeshiva University
500 West 185 Street
New York, NY
USA
lanave@yu.edu

3

Explicit constructions of Ricci solitons

PAUL BAIRD

Abstract

We describe methods for constructing explicit examples of Ricci solitons. An improved version of an anzatz of the author and L. Danielo for 3-dimensional solitons is given which we extend to higher dimensions. The soliton structure on the 4-dimensional geometry Nil_4 is analysed in detail, in particular its uniqueness is established and its relation to the natural Riemannian projection $\text{Nil}_4 \rightarrow \text{Nil}_3$ is discussed.

3.1 Introduction

The Ricci flow is the evolution equation:

$$\frac{\partial g}{\partial t} = -2\text{Ricci}\,(g) \tag{3.1}$$

for a time-dependent Riemannian metric $g = g(t)$ defined on an n-dimensional manifold M^n, subject to some initial condition $g(0) = g_0$. This equation has received a considerable amount of attention recently due to the work of G. Perelman [14] which has advanced Hamilton's program for resolving the geometrization conjecture. An essential part of this story are *Ricci solitons*, which correspond to fixed points, up to scaling and diffeomorphism. Specifically, if $g(t) = c(t)\psi_t^*(g_0)$ is a solution of (3.1) on some time interval $[0, \delta)$, where $c(t)$ is a family of positive scalars such that $c(0) = 1$ and ψ_t is a family of diffeomorphisms satisfying $\psi_0 = id$, then on calculating $g'(0)$ and applying (3.1) we obtain the equation:

$$2\text{Ricci}\,(g) + \mathcal{L}_E g + 2Ag = 0, \tag{3.2}$$

37

where we now write $g = g_0$, and where $2A = c'(0)$ and E is the vector field (called the *soliton flow*) determined at each $x \in M$ by $E(x) = \frac{d}{dt}\psi_t(x)|_{t=0}$. Conversely, on a complete manifold, any solution g_0 of (3.2) determines a solution of (3.1) of the form $g(t) = c(t)\psi_t^*(g_0)$ for some small time t (see [5]).

Equation (3.2) is known as the *soliton equation* and solutions play a fundamental role in the study of the Ricci flow: they occur as rescaled limits at singularity formation and as asymptotic limits of immortal solutions, that is solutions that exist for all future time [13]. Both of these limits are interpreted in terms of Cheeger-Gromov-Hamilton pointed convergence of Ricci flows [9]. In the case when $E = \text{grad } f$ is the gradient of a function, then the soliton is said to be of *gradient type*, in which case (3.2) becomes

$$\text{Ricci}\,(g) + \nabla^2 f + Ag = 0, \tag{3.3}$$

since $\mathcal{L}_E g = 2\nabla^2 f$ where $\nabla^2 f$ denotes the Hessian of f. A soliton defined by (3.2) is called *shrinking, steady* or *expanding* according as the constant A is negative, zero or positive, respectively.

At singularity formation, it is only gradient solitons of a specific type which occur, moreover, a gradient soliton is also a critical metric for Perelman's entropy functional [14]. Non-gradient solitons are important in the study of the long-term behaviour of solutions and it is only recently that the first examples were found [2, 13]. The existence of such solitons also has important consequences for the stability of the Ricci flow about non-Ricci flat metrics [8]; see also [7, 15] for the Ricci-flat case.

Clearly the soliton flow E is only defined up to addition of a Killing vector field and if it is itself Killing, then equation (3.2) becomes the equation for an Einstein metric. These solutions are referred to as the *trivial* Ricci solitons. In dimension 3, any compact soliton is trivial [10]. The greater flexibility in equation (3.2) makes the problem of finding non-trivial Ricci solitons both rich and challenging. Even in dimension 2, the complete solitons are not yet classified; see [5] Chapter 1, for a discussion.

One way to find explicit solutions to (3.2) is to suppose that the metric is sufficiently symmetric that the equations reduce to an ODE. The best-known example of such a construction is given by the *Bryant soliton*, which, in dimension two corresponds to the *Hamilton cigar*. In this case one considers a warped product metric of the form $g = dr^2 + F(r)^2 g_{S^n}$, where g_{S^n} is the standard metric on the n-sphere. One looks for a gradient soliton with soliton flow $E = \text{grad } f$ with f depending only on the parameter r. The system of equations to solve then becomes a dynamical system in $\mathbb{R}^2$. An orbit emanating from one of the fixed points gives a complete metric of positive sectional curvature [4] (see Section 3.4). This procedure was generalized by T. Ivey to construct solitons from doubly warped product metrics of the type $dr^2 + a(r)^2 g_1 + b(r)^2 g_2$, where g_1 and g_2 are metrics on a sphere and an Einstein manifold, resp. [11]. In Section 3.2, we make a further generalization which leads to a dynamical system. Here we no longer insist that the soliton be of gradient type and neither do we require that the soliton flow depend on a single parameter. The procedure picks up solitons of Sol-type, as well as a curious example on a surface. Although the soliton metric is neither complete nor compact, it factors and then extends by the addition of two points to a non-standard (complete) metric on the 2-sphere (Example 3.2.6).

In [2], the author and L. Danielo devised a procedure for constructing solitons in dimension 3, by reducing the number of variables to two, so that (3.2) becomes a system of equations on a Riemannian surface. It is important to note that the method does not consist simply of supposing quantities depend on two independent variables, rather, having solved the system of equations on the surface one is then required to solve an exterior differential equation in dimension 3 in order to obtain the 3-dimensional soliton metric. In particular, the metric itself and the soliton flow may depend on all three coordinates. In Section 3.3, we give an improved version of this ansatz and describe how the soliton structure on Nil_3 arises. Then in Section 3.4, we generalize the construction to more general dimensions.

On $\mathbb{R}^n$ there are the well-known *Gaussian solitons* with flow $E = \text{grad}(-\frac{A}{2}|x - a|^2)$, where $a \in \mathbb{R}^n$ is some arbitrary point and A is an arbitrary constant. This shows that a soliton structure may not be unique. However, it is not known whether there are other examples of Riemannian manifolds which support more than one soliton structure. One way to test this for a given manifold, is to solve the equation (3.2) directly. This requires sufficient symmetry of the metric in order that the system of equations is managable. In [2], this was done for the 3-dimensional geometries and uniqueness was established in all cases except $\mathbb{R}^3$, where one readily sees that the only examples are the Gaussian solitons. In Section 3.5, we do this for the 4-dimensional geometry Nil_4, to obtain the soliton structure whose existence was demonstrated by J. Lauret [12] by Lie group methods and is described more explicitly by J. Lott [13]. We further show uniqueness of this structure and discuss its relation to the one on Nil_3. The geometry Nil_4 naturally admits a harmonic Riemannian fibration to Nil_3, which allows us to relate their respective soliton structures.

The author thanks the *Agence Nationale de Recherche* project: *Flots et opérateurs géométriques* no: ANR-07-BLAN-0251-01 for financial support during the preparation of this paper.

3.2 Solitons from a dynamical system

Let us consider a 3-dimensional metric $g = g_{ij} dx^i dx^j$, which, in coordinates (x^1, x^2, x^3) is of diagonal form:

$$(g_{ij}) = \begin{pmatrix} 1 & 0 & 0 \\ 0 & a^2 & 0 \\ 0 & 0 & b^2 \end{pmatrix} \tag{3.4}$$

where the functions a, b both depend on x^1 only. As we shall see, the vector field E defining the soliton flow may also depend on the variable x^3. Then the components of the Ricci curvature are given by the formula:

$$R_{jk} = \partial_l \Gamma^l_{jk} - \partial_j \Gamma^l_{lk} + \Gamma^l_{lm} \Gamma^m_{jk} - \Gamma^l_{jm} \Gamma^m_{lk},$$

and we find that the Ricci curvature is also diagonal with components:

$$R_{11} = -(u' + v' + u^2 + v^2), \quad R_{22} = -a^2(v' + uv + v^2), \quad R_{33} = -b^2(u' + u^2 + uv).$$

In order to obtain a dynamical system, we suppose that the soliton flow E in (3.2) has the form:

$$E = X + fU \tag{3.5}$$

where $U = \frac{1}{b} \partial_3$ and f is some function. We suppose further that $X = \operatorname{grad} \beta$, where $\beta = \beta(x^1, x^2)$. This is not the same as supposing that E is of gradient type, but rather that it is of gradient type modulo a component in the direction of the vector field U. The reason that we make this decomposition is that, by an application of the Poincaré Lemma, it is possible to eliminate the function f from (3.2), so reducing the number of unknown parameters by one. A further useful consequence is to essentially fix a *gauge* for E, since in general, the soliton flow is defined only up to addition of a Killing vector field. We find that orbits of our dynamical system are independent of any freedom that remains in the choice of E. The following theorem is proved in [1]

Theorem 3.2.1 *Let A and k be constants and consider the 3-dimensional autonomous dynamical system:*

$$\begin{cases} u' = u\rho - k \\ v' = v\rho + A \\ \rho' = u^2 + v^2 - A. \end{cases} \tag{3.6}$$

Then each orbit $t \mapsto (u(t), v(t), \rho(t))$ determines a Ricci soliton metric of the form $g = (dx^1)^2 + a(t)^2 (dx^2)^2 + b(t)^2 (dx^3)^2$ via the correspondence: $t = x^1$, $u = b'/b$, $v = a'/a$. The soliton is shrinking, steady or expanding, according as $A < 0, = 0, > 0$, respectively. *Up to addition of a Killing vector field, the soliton flow is given by:*

$$E = (\rho + u + v)\frac{\partial}{\partial x^1} - (k + A)x^3 \frac{\partial}{\partial x^3} ; \tag{3.7}$$

it is of gradient type if and only if either $k + A = 0$, or $b' = 0$ $(u = 0)$.

A first observation is that the constant A in (3.6) can be normalized by permitting a homothetic change of the metric g. More precisely, set $s = ct$ (c constant), so that

$$g = \frac{ds^2}{c^2} + a(t)^2(dx^2)^2 + b(t)^2(dx^3)^2$$

$$= \frac{1}{c^2}\left(ds^2 + \tilde{a}(s)^2(dx^2)^2 + \tilde{b}(s)^2(dx^3)^2\right)$$

where we have written $\tilde{a}(s) = ca(s/c)$, $\tilde{b}(s) = cb(s/c)$. If we define $\tilde{u}(s) = \tilde{b}'(s)/\tilde{b}(s) = u(t)/c$, $\tilde{v}(s) = \tilde{a}'(s)/\tilde{a}(s) = v(t)/c$. Then

$$\tilde{u}'(s) = \frac{1}{c^2}u'(t) = \frac{1}{c^2}(u\rho - k) = \tilde{u}\tilde{\rho} - \frac{k}{c^2}$$

where we have defined $\tilde{\rho}(s) = \rho(s/c)/c$. Similarly for the other derivatives and the dynamical system (3.6) now becomes:

$$\begin{cases} \tilde{u}'(s) = \tilde{u}\tilde{\rho} - \frac{k}{c^2} \\ \tilde{v}'(s) = \tilde{v}\tilde{\rho} + \frac{A}{c^2} \\ \tilde{\rho}'(s) = \tilde{u}^2 + \tilde{v}^2 - \frac{A}{c^2} \end{cases} \tag{3.8}$$

By choosing c appropriately, we can now normalise the constant A to be $-2, 0$ or 2.

We now show how a number of solitons metrics, some trivial, some non-trivial, arise as orbits of (3.6).

Example 3.2.2 (The geometry Sol) For $A > 0$, the equilibrium points are given by

$$\left(\frac{\pm k\sqrt{A}}{\sqrt{k^2 + A^2}}, \frac{\mp A\sqrt{A}}{\sqrt{k^2 + A^2}}, \pm\frac{\sqrt{k^2 + A^2}}{\sqrt{A}}\right).$$

This gives the metric

$$\begin{pmatrix} 1 & 0 & 0 \\ 0 & e^{2A\sqrt{\frac{A}{k^2 + A^2}}t} & 0 \\ 0 & 0 & e^{-2k\sqrt{\frac{A}{k^2 + A^2}}t} \end{pmatrix}$$

But after normalisation of the constant A, we may suppose that

$$A\sqrt{\frac{A}{k^2 + A^2}} = 1.$$

Now write $\lambda = k/A$ to give the form:

$$\begin{pmatrix} 1 & 0 & 0 \\ 0 & e^{2t} & 0 \\ 0 & 0 & e^{-2\lambda t} \end{pmatrix}$$

where λ is an arbitrary parameter. In the case when $\lambda = 1$ this gives the metric for the geometry Sol, in particular, Sol is one of a 1-parameter family of soliton metrics.

Example 3.2.3 (Constant curvature examples) The sphere S^3 arises by setting $A = -2$, $k = 2$, $a = \sin t$ and $b = \cos t$. Then $u = -\tan t$, $v = \cot t$ and the corresponding orbit of (3.6) is given by setting $\rho = \frac{1}{\sin t \cos t} - 2 \cot t$.

In a similar fashion, hyperbolic 3-space occurs by setting $A = 2$, $k = -2$, $a = \sinh t$ and $b = \cosh t$. Then $u = \tanh t$, $v = \coth t$ and the corresponding orbit occurs by setting $\rho = -\frac{1}{\sinh t \cosh t} - 2 \tanh t$.

Note that in both cases, the orbits are not defined for all time. Also, in both cases, with reference to Theorem 3.2.1, we have $k + A = 0$.

Example 3.2.4 (Product examples) Examples of the form $\Sigma \times \mathbb{R}$ occur when $b \equiv 1$. Then $u \equiv 0$ and we obtain the 2-dimensional dynamical system:

$$\begin{cases} v' = v\rho + A \\ \rho' = v^2 - A. \end{cases}$$

But this is precisely the system that occurs in the construction of a 2-dimensional Ricci soliton with metric of the form: $h = (dx^1)^2 + a(x^1)^2(dx^2)^2$. For example, Hamilton's 2-dimensional cigar is given by the metric: $h = (dx^1)^2 + \tanh^2 t(dx^2)^2$. Then it is easily seen that the above equations are satisfied with $A = 0$. This example is a complete steady gradient soliton; for more details, see for example [6].

Example 3.2.5 (Warped product examples) We consider the special case when $a \equiv b$, which can be considered as a warped product of the real line with standard Euclidean 2-space. Now our system can be viewed as the sub-system of the dynamical system (3.6) given by the orbits passing through the axis $u = v = 0$ when $A + k = 0$. It is easily seen that the plane $u = v$ is an invariant subspace, so the orbits remain in this plane. This then leads to the 2-dimensional dynamical system:

$$\begin{cases} u' = u\rho + A \\ \rho' = 2u^2 - A. \end{cases}$$

By eliminating ρ, we see that u is given as a solution to the second order ODE:

$$u''u - (u')^2 + Au' + Au^2 - 2u^4 = 0.$$

A particular solution when $A = 0$ is given by $u = \pm\frac{1}{\sqrt{2}}t^{-1}$, which yields the metric $g = dt^2 + t^{\pm\sqrt{2}}((dx^2)^2 + (dx^3)^2)$. This singular metric of negative scalar curvature was also noticed in [2]. It corresponds to a steady gradient 3-dimensional soliton. If we don't insist on completeness, this example provides an answer to Problem 1.88 of [5]; by the expressions for the components of the Ricci curvature, we see that the Ricci tensor is positive whatever sign is taken in the expression for u, as is required – cf. the discussion in [5].

Example 3.2.6 (Non-standard structure over S^2) Let us make the orthogonal substitution:

$$\xi = \frac{Au + kv}{\sqrt{k^2 + A^2}}, \qquad \eta = \frac{-ku + Av}{\sqrt{k^2 + A^2}}$$

which puts (3.6) into the form:

$$\begin{cases} \xi' = \xi\rho \\ \eta' = \eta\rho + \sqrt{k^2 + A^2} \\ \rho' = \xi^2 + \eta^2 - A. \end{cases} \tag{3.9}$$

We study the orbit with initial condition $(\xi(0), \eta(0), \rho(0)) = (0, 0, 0)$, so that $\xi(t) = 0$ for all t, and we now have the reduced system:

$$\begin{cases} \eta' = \eta\rho + \sqrt{k^2 + A^2} \\ \rho' = \eta^2 - A. \end{cases} \tag{3.10}$$

If $A > 0$, the fixed points are given by $\left(\pm\sqrt{A}, \mp\sqrt{\dfrac{k^2 + A^2}{A}}\right)$, otherwise there are none unless $A = k = 0$ in which case the whole axis $\eta = 0$ consists of fixed points. We study the case $A = 0$, $k \neq 0$; for the other cases, see [1].

The dynamical system (3.10) now takes the form

$$\begin{cases} \eta' = \eta\rho + k \\ \rho' = \eta^2. \end{cases} \tag{3.11}$$

On replacing η with $-\eta$, it is no loss of generality to suppose that $k > 0$. We consider the orbit with initial condition $(\eta(0), \rho(0)) = (0, 0)$. Then there exists a $T > 0$ such that $\eta(t), \rho(t) \to \infty$ as $t \to T^-$. In [1] it is shown that this orbit determines a steady soliton metric g of non-constant curvature on the surface $(0, 2T) \times \mathbb{R}$. The soliton is of non-gradient type and its flow is given by the vector field

$$E = (\eta - \rho)\frac{\partial}{\partial r} - kx^3\frac{\partial}{\partial x^3} \,,$$

where $r = T - t$. In fact $\lim_{r \to 0^+}(\eta(r) - \rho(r)) = \lim_{r \to 2T^-}(\eta(r) - \rho(r)) = 0$. The metric g factors to a metric on the cylinder $(0, 2T) \times S^1$ and then by the *cylinder to ball rule*, extends to a C^2 metric on the sphere S^2.

3.3 Reduction of the equations to a 2-dimensional system

In [2], an ansatz is given for constructing a soliton metric on a 3-manifold M^3 from data on a surface N^2. The basis of this ansatz is the relation between objects on N^2 and M^3 given that there exists a semi-conformal mapping $\varphi : M^3 \to N^2$. In this section we give an improved version of this ansatz which highlights how the soliton equation on M^3 corresponds to a "deformed" soliton equation on N^2.

A smooth map $\varphi : (M^m, g) \to (N^n, h)$ between manifolds M, N with respective Riemannian metrics g, h is said to be *semi-conformal* if, for each $x \in M$ where $d\varphi_x \neq 0$, the restriction $d\varphi_x|_{\mathcal{H}_x} : \mathcal{H}_x \to T_{\varphi(x)}N$ is conformal and surjective, where $\mathcal{H}_x = (\ker d\varphi_x)^{\perp}$. Thus there exists a number $\lambda = \lambda(x) > 0$ such that $\varphi^*h(X, Y) = \lambda(x)^2 g(X, Y)$ for all $X, Y \in \mathcal{H}_x$. Setting $\lambda = 0$ at points $x \in M$ where $d\varphi_x = 0$, we obtain a continuous function $\lambda : M \to \mathbb{R}(\geq 0)$ called the *dilation of* φ; it has the property that $\lambda^2 = \frac{1}{n}||d\varphi||^2$ is smooth, where $||d\varphi||$ denotes the Hilbert-Schmidt norm of the derivative at each point. The fundamental equation of a semi-conformal submersion relates the tension field, the dilation and the mean curvature μ of fibres by the formula [3]:

$$\tau(\varphi) = -(n - 2)d\varphi(\text{grad} \ln \lambda) - (m - n)d\varphi(\mu).$$

If we now suppose that $\dim M = \dim N + 1$ with both M and N oriented (or simply, the fibres of φ are oriented) and that φ is submersive, then it follows that the metric g has the form

$$g = \frac{\varphi^*h}{\lambda^2} + \theta^2$$

for some 1-form θ on M, non-vanishing on $\ker d\varphi$.

Let $\varphi : (M^3, g) \to (N^2, h)$ be a semi-conformal submersion. Then there are three quantities that come into play when setting up a correspondence as described in the opening paragraph: the dilation $\lambda : M^3 \to \mathbb{R}$ (> 0); the mean-curvature of the fibres $\mu : M \to TM$; the integrability tensor of the horizontal distribution $I : TM \times TM \to TM$ given by $I(X, Y) = \mathcal{V}[\mathcal{H}X, \mathcal{H}Y]$, where $\mathcal{H}$ and $\mathcal{V}$ are orthogonal projection onto the horizontal and vertical spaces, respectively. We impose further restrictions by supposing that μ is the gradient of a function and that the above three quantities are basic: $\lambda = \overline{\lambda} \circ \varphi$, $\mu = \operatorname{grad} \rho$ with $\rho = \overline{\rho} \circ \varphi$, $\psi := \|I\|^2 = \sum_{a,b} \|I(e_a, e_b)\|^2 = \overline{\psi} \circ \varphi$, where $\{e_a\}$ is an orthonormal basis for $T_x M$ at each $x \in M$ and where we place a "bar" over corresponding quantities on N^2.

The two most important objects to express on M^3 are its metric g and its Ricci curvature Ricci (g). Then, as above,

$$g = \frac{\varphi^* h}{\lambda^2} + \theta^2,$$

where $\theta = g(U, \cdot)$ with U a unit tangent to the fibres of φ. For the Ricci curvature, we obtain [2]:

$$\operatorname{Ricci}(g) = \left\{ \lambda^2 K^N + \Delta \ln \lambda + \mu(\ln \lambda) \right\} (g - \theta^2) - \frac{1}{4} \|I\|^2 g + \frac{1}{2} \mathcal{L}_\mu g$$
$$- (\mu^\flat + \mathrm{d}^\mathcal{V} \ln \lambda)^2 - (\mathrm{d}^\mathcal{V} \ln \lambda)^2 + \Delta\theta \odot \theta, \qquad (3.12)$$

where K^N is the Gauss curvature of N, $\Delta\theta = (\mathrm{dd}^* + \mathrm{d}^*\mathrm{d})\theta$ is the Laplacian on forms and $\odot$ denotes the symmetrized tensor product.

We now consider the soliton flow, which, as in Section 3.2, we suppose has a decomposition of the form:

$$E = X + fU,$$

where $g(X, U) = 0$ and f is to be determined. In essence, one of the components of the soliton equation becomes $\mathrm{d}(f\rho) = \ldots$ where the right hand-side is independent of f, so that the existence of f becomes an application of the Poincaré Lemma (see [2] for details). We then make the assumption that X is a gradient with basic potential: $X = \operatorname{grad} \ln \nu$, where $\nu = \overline{\nu} \circ \varphi$.

It is a straightforward calculation to show that $\Omega := \mathrm{d}\theta = -g(U, I) - \mu^\flat \wedge \theta$ and, under the assumption that μ is a gradient, the 2-form $\widetilde{\Omega} = -g(U, I)$ is always basic (cf. [2] Lemma 3.1). Specifically, $\widetilde{\Omega} = \varphi^* \overline{\Omega}$, where $\overline{\Omega} = (\sqrt{2\overline{\psi}}/\overline{\lambda}^2)\mu^N$ and where μ^N is the volume form on N. Thus in order to construct θ we are required to solve the exterior differential equation:

$$\mathrm{d}\theta + \mu^\flat \wedge \theta = \widetilde{\Omega}.$$

We are now ready to state an improved version of the ansatz given in [2].

Theorem 3.3.1 *Let (N^2, h) be a Riemannian surface and let $\overline{\rho}, \overline{v}, \overline{\zeta} : N \to \mathbb{R}$ satisfy the following system of equations:*

$$
\begin{cases}
\text{(i)} & \text{Ricci}^N + \nabla^2 \ln \overline{v} - (d \ln \overline{\rho})^2 + (A - \overline{\zeta}^2)h = 0 \\
\text{(ii)} & \Delta \ln \overline{\rho} - h(\text{grad} \ln \overline{\rho}, \text{grad} \ln \overline{v}) + \overline{\zeta}^2 + B = 0 \\
\text{(iii)} & \Delta \ln(\overline{\rho}^2 \overline{v} \overline{\zeta}^{-1}) + |\text{grad} \ln \overline{\rho}|^2 - |\text{grad} \ln \overline{\zeta}|^2 \\
& \quad + h(\text{grad} \ln \overline{\zeta}, \text{grad} \ln \overline{v}) + \overline{\zeta}^2 + A = 0
\end{cases}
\tag{3.13}
$$

where A and B are constants and where equation (iii) *is vacuous if $\overline{\zeta} \equiv 0$. Set $M = N \times (-\delta, \delta)$ for some $\delta > 0$, and let $\varphi : M \to N$ be the canonical projection. Let $\overline{\Omega} = \sqrt{2}\,\overline{\zeta}\,\mu^N$, $\widetilde{\Omega} = \varphi^* \overline{\Omega}$, $\rho = \overline{\rho} \circ \varphi$. Let θ be a 1-form satisfying*

$$
d\theta + d \ln \rho \wedge \theta = \widetilde{\Omega}, \tag{3.14}
$$

which is non-vanishing on $\ker d\varphi$. *Write $g = \varphi^* h + \theta^2$. Then (M^3, g) is a Ricci soliton.*

Proof: By [2], it is sufficient to prove the equivalence of the above system with the following set of equations:

$$
\begin{cases}
\text{(i)}'\text{(a)} & K^N + \frac{1}{2}\left(\Delta^N \ln(\overline{\lambda}^2 \overline{v}) - |\text{grad} \ln \overline{\rho}|^2\right) + \frac{A - \overline{\psi}}{\overline{\lambda}^2} = 0 \\
\text{(i)}'\text{(b)} & \nabla^2 \ln \overline{v} + 2d \ln \overline{\lambda} \odot d \ln \overline{v} - (d \ln \overline{\rho})^2 = \alpha h \quad (\text{some } \alpha : N \to \mathbb{R}) \\
\text{(ii)}' & \overline{\lambda}^2\left\{\Delta^N \ln \overline{\rho} - h(\text{grad} \ln \overline{\rho}, \text{grad} \ln \overline{v}) + \frac{\overline{\psi} + A}{\overline{\lambda}^2}\right\} = \text{const.} \\
\text{(iii)}' & \Delta^N \ln(\overline{\rho}^2 \overline{v} \overline{\psi}^{-1/2}) + |\text{grad} \ln \overline{\rho}|^2 - \frac{1}{4}|\text{grad} \ln \overline{\psi}|^2 \\
& \quad + \frac{1}{2}h(\text{grad} \ln \overline{\psi}, \text{grad} \ln \overline{v}) + \frac{\overline{\psi} + A}{\overline{\lambda}^2} = 0,
\end{cases}
\tag{3.15}
$$

where (iii)$'$ is vacuous whenever $\overline{\psi} \equiv 0$. However, this system is conformally invariant (see [2]), and so we can replace h by $\widetilde{h} = \overline{\lambda}^{-2} h$ and so, without loss of generality, we can suppose that $\overline{\lambda} \equiv 1$. Note that by doing this we break the conformal invariance. We also replace $\overline{\psi} \geq 0$ by $\overline{\zeta}$ determined by $\overline{\zeta}^2 = \overline{\psi}$ (now there is no restriction on the sign of $\overline{\zeta}$). We claim that (i) is equivalent to (i)$'$(a) and (i)$'$(b).

Since Ricci$^N = K^N h$, then (i)$\Rightarrow$(i)$'$(b) with $\alpha = \overline{\zeta}^2 - A - K^N$. Now the trace of (i) implies that

$$
2K^N + \delta \ln \overline{v} - |\text{grad} \ln \overline{\rho}|^2 + 2A - 2\overline{\zeta}
$$

which is equivalent to (i)$'$(a). Thus (i) implies (i)$'$(a) and (b). Conversely, given (i)$'$(a) and (b), taking the trace of (i)$'$(b) gives

$$
2\alpha = \Delta \ln \overline{v} - |\text{grad} \ln \overline{\rho}|^2,
$$

which by (i)′(a) equals $-2K^N - 2A + 2\bar{\zeta}^2$, so that

$$\nabla^2 \ln \bar{v} - (d \ln \bar{\rho})^2 + K^N h + AH - \bar{\zeta}^2 h = 0,$$

which is precisely (i). □

Example 3.3.2 (Case of integrable horizontal distribution) If $\bar{\zeta} \equiv 0$, then the horizontal distribution of φ is integrable and equation (3.13)(iii) becomes vacuous (cf. [2]). We then have the following coupled system to solve:

$$\begin{cases} \text{(i)} & \text{Ricci}^N + \nabla^2 \ln \bar{v} - (d \ln \bar{\rho})^2 + Ah = 0 \\ \text{(ii)} & \Delta \ln \bar{\rho} - h(\text{grad} \ln \bar{\rho}, \text{grad} \ln \bar{v}) + B = 0. \end{cases}$$

For example, the dynamical system of Theorem 3.2.1 can be seen to arise this way.

Example 3.3.3 (Case of minimal fibres–the geometry Nil) The fibres of φ are minimal if and only if $\bar{\rho} = \text{const}$. In this case equation (ii) implies that $\bar{\zeta}^2 = -B$ is constant and (i) becomes:

$$\text{Ricci}^N + \nabla^2 \ln \bar{v} + (A - B)h = 0,$$

which is the equation for a gradient soliton on the surface (N^2, h). If $\bar{\zeta} = 0$, then the horizontal distribution is integrable and we locally have a product structure $M^3 = \Sigma \times \mathbb{R}$, where Σ is a 2-dimensional gradient soliton.

If on the other hand $\bar{\zeta} \neq 0$, then equation (iii) implies that

$$\Delta \ln \bar{v} + \bar{\zeta}^2 + A = 0,$$

which combined with the trace of (i) gives the identity:

$$2K^N + A - 3\bar{\zeta}^2 = 0,$$

so that in particular the curvature K^N must be constant. If we take $N = \mathbb{R}^2$ with its canonical metric and coordinates (x, y) and set $\bar{\zeta} = 1/\sqrt{2}$, then, applying the construction of Theorem 3.3.1, we obtain $\widetilde{\Omega} = dx \wedge dy$. To find θ we are required to solve

$$d\theta = dx \wedge dy.$$

so that, up to a diffeomorphism, we can take $\theta = xdy + dz$ (recalling that θ must not vanish on vertical vectors). This then gives the soliton metric:

$$g = dx^2 + dy^2 + (xdy + dz)^2,$$

which is the metric of the geometry Nil. It is easy to calculate the soliton flow, which is given by

$$E = -x\frac{\partial}{\partial x} - y\frac{\partial}{\partial y} - 2z\frac{\partial}{\partial z}.$$

This is not a gradient with respect to the metric g [2].

3.4 Higher dimensional Ricci solitons via projection

In general dimensions, the problem of understanding the soliton equations in terms of the parameters of a semi-conformal submersion $\varphi : M^m \to N^n$ become more difficult. This is because the fundamental tensors of a submersion come into play when expressing the Ricci curvature on M^m in terms of that on N^n, see [3]. However, provided the fibres have dimension 1 and φ is harmonic, i.e. a harmonic morphism, then we still obtain a manageable expression.

Proposition 3.4.1 *Let $\varphi : M^{n+1} \to N^n$ be a submersive harmonic morphism, then*

$$
\begin{aligned}
\text{Ricci}^M = {} & \varphi^* \text{Ricci}^N + \Delta \ln \lambda \, g^{\mathcal{H}} - \frac{n(n-2)}{2} d \ln \lambda^2 \\
& + \left(n \, d(U(\ln \lambda)) + d^* \Omega + n(n-2) U(\ln \lambda) d \ln \lambda \right) \odot \theta \\
& - \left(\frac{n^2}{2} U(\ln \lambda)^2 + \frac{1}{4} ||\Omega||^2 \right) \theta^2 - \frac{1}{2} \Omega(e_i, \cdot) \Omega(e_i, \cdot),
\end{aligned}
$$

where $\{e_i\}$ is an orthonormal basis of the horizontal space $(\ker d\varphi)^{\perp}$ and $g^{\mathcal{H}} = \varphi^* h / \lambda^2$ is the horizontal part of the metric.

Proof: As before, let U be a unit vector field tangent to the fibres of φ and let $\theta = g(U, \cdot)$. If one rescales: $V = \lambda^{n-2} U$, $\widetilde{\theta} = \lambda^{-n+2} \theta$, $\widetilde{\Omega} = d\widetilde{\theta}$, then in [3], the Ricci curvature is computed on components as follows; here we suppose that X, Y are orthogonal to U:

$$
\begin{aligned}
\text{Ricci}^M(U, U) = {} & -(n-2)\Delta \ln \lambda + 2(n-2)U(U(\ln \lambda)) \\
& - n(n-1)||\mathcal{V}\text{grad} \ln \lambda||^2 + \tfrac{1}{4} \lambda^{2n-4} ||\widetilde{\Omega}||^2 \\
\text{Ricci}^M(X, U) = {} & (n-1)X(U(\ln \lambda)) \\
& + \tfrac{1}{2} \lambda^{n-2} \left\{ d^* \widetilde{\Omega}(X) - 2(n-2)\widetilde{\Omega}(\text{grad} \ln \lambda, X) \right\} \\
\text{Ricci}^M(X, Y) = {} & \text{Ricci}^N(d\varphi(X), d\varphi(Y)) + \langle X, Y \rangle \Delta \ln \lambda \\
& - (n-1)(n-2)X(\ln \lambda)Y(\ln \lambda) - \tfrac{1}{2} \lambda^{2n-4} \langle i_X \widetilde{\Omega}, i_Y \widetilde{\Omega} \rangle,
\end{aligned}
$$

where i_X denotes contraction with respect to the vector X. Now

$$
\Omega = \lambda^{n-2} \widetilde{\Omega} + (n-2) d \ln \lambda \wedge \theta,
$$

from which, after some calculation, we obtain

$$
\begin{aligned}
d^* \Omega = {} & \lambda^{n-2} d^* \widetilde{\Omega} - (n-2)\lambda^{n-2} \widetilde{\Omega}(\text{grad} \ln \lambda, \cdot) \\
& + (n-2) \left\{ \left(-\Delta \ln \lambda - nU(\ln \lambda)^2 + (n-2)||\mathcal{H}\text{grad} \ln \lambda||^2 \right) \theta \right. \\
& \left. + d(U(\ln \lambda)) \right\}.
\end{aligned}
$$

On substituting this and evaluating on different combinations of horizontal and vertical vectors, the formula of the proposition follows. $\qquad\square$

We are now able to obtain the soliton equations in terms of the parameters of a semi-conformal submersion as expressed by the following theorem.

Theorem 3.4.2 *Let* $\varphi : M^{n+1} \to N^n$ *be a submersive harmonic morphism. Then* M^{n+1} *admits a soliton structure with flow* $E = X + fU$ *where* $g(X, U) = 0$ *and with constant A if and only if:*

$$
\begin{aligned}
0 = {}& \varphi^*\text{Ricci}^{\,N} + \left(\Delta \ln \lambda - fU(\ln \lambda)\right)g^{\mathcal{H}} - \frac{n(n-2)}{2}\mathrm{d}\ln\lambda^2 \\
& + \left\{ n\mathrm{d}(U(\ln\lambda)) + \mathrm{d}^*\Omega + n(n-2)U(\ln\lambda)\mathrm{d}\ln\lambda + \mathrm{d}f - (n-2)f\mathrm{d}\ln\lambda \right\} \odot \theta \\
& + \left\{ (n-2)fU(\ln\lambda) - \frac{n^2}{2}U(\ln\lambda)^2 - \frac{1}{4}\|\Omega\|^2 \right\} \theta^2 \\
& - \frac{1}{2}\Omega(e_i, \cdot)\Omega(e_i, \cdot) + \frac{1}{2}\mathcal{L}_X g + Ag \,,
\end{aligned}
$$

where $\{e_i\}$ *is an orthonormal basis for* $(\ker \mathrm{d}\varphi)^\perp$.

In order to obtain the above formula, it suffices to note that

$$
\mathcal{L}_{fU}g = -2fU(\ln\lambda)g^{\mathcal{H}} + 2(\mathrm{d}f - (n-2)f\mathrm{d}\ln\lambda)\odot\theta + 2(n-2)fU(\ln\lambda)\theta^2,
$$

and to then apply the soliton equation (3.2).

Example 3.4.3 (The Bryant solitons) We consider $M^{n+1} = N^n \times J$ endowed with a metric of warped propduct type: $g = (h/\lambda^2) + \mathrm{d}t^2$, where J is an open interval in $\mathbb{R}$ with its coordinate t and where $\lambda = \lambda(t)$. We take φ to be the canonical projection $\varphi : M \to N$. Then the fibres are geodesic and φ is both semi-conformal and harmonic. Note also that the horizontal distribution is integrable, so that Ω vanishes. We will suppose that the horizontal component X of the soliton flow E vanishes and that N is Einstein, with $\text{Ricci}^{\,N} = ah$, so that $\varphi^*\text{Ricci}^{\,N} = a\lambda^2 g^{\mathcal{H}}$. Then the equation of the above theorem is equivalent to the pair of equations

$$
\begin{cases}
0 & = \; a\lambda^2 + \Delta\ln\lambda - fU(\ln\lambda) + A \\
0 & = \; U(f) + nU(U(\ln\lambda)) - nU(\ln\lambda)^2 + A \,.
\end{cases}
$$

Now suppose that t denotes a unit speed parameter along the fibres, so that $U = \partial/\partial t$ and $\theta = \mathrm{d}t$, then a necessary consequence of the equations is that $f = f(t)$. There are product solutions given by $\lambda = $const. If, on the other hand, we suppose that λ is non-constant and work on a neighbourhood where $\lambda' \neq 0$, then, on letting $\{e_a\}$ denote a local orthonormal frame, we obtain

$$\Delta \ln \lambda = \text{Tr}\, \nabla^2 \ln \lambda = d \ln \lambda (\nabla_{e_a} e_a) + U(U(\ln \lambda) = -n((\ln \lambda)')^2 + (\ln \lambda)''.$$

We thereby obtain the system:

$$\begin{cases} 0 = \lambda^2 a + (\ln \lambda)'' - n((\ln \lambda)')^2 - f(\ln \lambda)' + A \\ 0 = f' + n(\ln \lambda)'' - n((\ln \lambda)')^2 + A \end{cases}$$

One can eliminate f from these equations to obtain a 3rd order ODE in λ, solutions of which correspond to soliton metrics (see [2] for the three-dimensional case). The Bryant solitons arise by taking $A = 0$ and making the substitutions: $\rho = 1/\lambda$, $x = \rho'$, $y = \rho f + n\rho'$, $ds = dt/\rho$, thus leading to a system of the form: $x'(s) = F(x, y)$, $y'(s) = G(x, y)$ for appropriate F and G. In particular, on taking N to be the n-sphere with its canonical metric, one obtains a solution corresponding to a complete rotationally symmetric steady soliton [4].

3.5 The 4-dimensional geometry Nil_4

If a metric has sufficient symmetry, it may be possible to solve the soliton equations directly. This was done in [2] in order to show non-existence of any soliton structure on the geometry $\widetilde{SL}_2(\mathbb{R})$ and uniqueness on the other 3-dimensional geometries. Here we perform the calculations for the 4-dimensional geometry Nil_4 and so produce an explicit coordinate expression for the soliton flow as well as showing its uniqueness.

The geometry Nil_4 can be identified with $\mathbb{R}^4$ endowed with the metric

$$g = dx_1{}^2 + dx_3{}^2 + (dx_2 + x_1 dx_3)^2 + \left(dx_4 + x_1 dx_2 + \frac{x_1{}^2}{2} dx_3 \right)^2.$$

This can be calculated from its characterisation as a left-invariant metric with respect to the group structure of Nil_4. There is a natural harmonic Riemannian submersion $(x_1, x_2, x_3, x_4) \to (x_1, x_2, x_3)$ onto $(\mathbb{R}^3, h = dx_1{}^2 + dx_3{}^2 + (dx_2 + x_1 dx_3)^2) = \text{Nil}_3$.

The components of the inverse matrix $(g^{ij}) = (g_{ij})^{-1}$ are given by

$$\begin{pmatrix} 1 & 0 & 0 & 0 \\ 0 & 1 + x_1{}^2 & -x_1 & -x_1 \left(1 + \frac{x_1{}^2}{2} \right) \\ 0 & -x_1 & 1 & \frac{x_1{}^2}{2} \\ 0 & -x_1 \left(1 + \frac{x_1{}^2}{2} \right) & \frac{x_1{}^2}{2} & \left(1 + \frac{x_1{}^2}{2} \right)^2 \end{pmatrix}$$

from which we can compute the non-zero Christofell symbols:

$$\Gamma_{22}^1 = -x_1, \quad \Gamma_{23}^1 = -\tfrac{1}{2}\left(1 + \tfrac{3x_1^2}{2}\right), \quad \Gamma_{24}^1 = -\tfrac{1}{2}, \quad \Gamma_{33}^1 = -x_1\left(1 + \tfrac{x_1^2}{2}\right),$$

$$\Gamma_{34}^1 = -\tfrac{x_1}{2}, \quad \Gamma_{13}^2 = \tfrac{1}{2}\left(1 - \tfrac{x_1^2}{2}\right), \qquad \Gamma_{14}^2 = \tfrac{1}{2}, \qquad \Gamma_{12}^3 = \tfrac{1}{2},$$

$$\Gamma_{13}^3 = \tfrac{x_1}{2}, \qquad \Gamma_{12}^4 = \tfrac{1}{2}\left(1 - \tfrac{x_1^2}{2}\right), \qquad \Gamma_{14}^4 = -\tfrac{x_1}{2},$$

Then we compute the Ricci curvature, which is given by:

$$\text{Ricci} = -dx_1{}^2 - \frac{1}{2}dx_3{}^2 + \frac{1}{2}\left(x_1 dx_2 + \frac{x_1{}^2}{2}dx_3 + dx_4\right)^2.$$

We now introduce the soliton flow $E = \alpha \partial_1 + \beta \partial_2 + \gamma \partial_3 + f \partial_4$, compute $\mathcal{L}_E g$ and substitute into the soliton equation (3.2). We omit the details of this latter calculation. When we equate the different coefficients to zero, we obtain the following set of equations to solve:

$dx_1{}^2:$ $\quad -2 + 2\partial_1\alpha + 2A = 0$

$dx_1 dx_2:$ $\quad 2\partial_2\alpha + 2(1 + x_1{}^2)\partial_1\beta + 2x_1\left(1 + \tfrac{x_1^2}{2}\right)\partial_1\gamma + 2x_1\partial_1 f = 0$

$dx_1 dx_3:$ $\quad 2\partial_3\alpha + 2x_1\left(1 + \tfrac{x_1^2}{2}\right)\partial_1\beta + 2\left(1 + \tfrac{x_1^2}{2}\right)^2\partial_1\gamma + x_1{}^2\partial_1 f = 0$

$dx_1 dx_4:$ $\quad 2\partial_4\alpha + 2x_1\partial_1\beta + x_1{}^2\partial_1\gamma + 2\partial_1 f = 0$

$dx_2{}^2:$ $\quad x_1{}^2 + 2x_1\alpha + 2(1 + x_1{}^2)\partial_2\beta + 2x_1\left(1 + \tfrac{x_1^2}{2}\right)\partial_2\gamma$

$\qquad\qquad + 2x_1\partial_2 f + 2(1 + x_1{}^2)A = 0$

$dx_2 dx_3:$ $\quad x_1{}^3 + (2 + 3x_1{}^2)\alpha + 2(1 + x_1{}^2)\partial_3\beta + 2x_1\left(1 + \tfrac{x_1^2}{2}\right)\partial_3\gamma + 2x_1\partial_3 f$

$\qquad\qquad + 2x_1\left(1 + \tfrac{x_1^2}{2}\right)\partial_2\beta + 2\left(1 + \tfrac{x_1^2}{2}\right)^2\partial_2\gamma + x_1{}^2\partial_2 f + 2A(2x_1 + x_1{}^3) = 0$

$dx_2 dx_4:$ $\quad 2x_1 + 2\alpha + 2(1 + x_1{}^2)\partial_4\beta + 2x_1\left(1 + \tfrac{x_1^2}{2}\right)\partial_4\gamma + 2x_1\partial_4 f$

$\qquad\qquad + 2x_1\partial_2\beta + x_1{}^2\partial_2\gamma + 2\partial_2 f + 4x_1 A = 0$

$dx_3{}^2:$ $\quad -1 + \tfrac{x_1^4}{4} + 2x_1\left(1 + \tfrac{x_1^2}{2}\right)\alpha + 2x_1\left(1 + \tfrac{x_1^2}{2}\right)\partial_3\beta$

$\qquad\qquad + 2\left(1 + \tfrac{x_1^2}{2}\right)^2\partial_3\gamma + x_1{}^2\partial_3 f + 2A\left(1 + \tfrac{x_1^2}{2}\right)^2 = 0$

$dx_3 dx_4:$ $\quad x_1{}^2 + 2x_1\alpha + 2x_1\left(1 + \tfrac{x_1^2}{2}\right)\partial_4\beta + 2\left(1 + \tfrac{x_1^2}{2}\right)\partial_4\gamma + x_1{}^2\partial_4 f$

$\qquad\qquad + 2x_1\partial_3\beta + x_1{}^2\partial_3\gamma + 2\partial_3 f + 2Ax_1{}^2 = 0$

$dx_4{}^2:$ $\quad 1 + 2x_1\partial_4\beta + x_1{}^2\partial_4\gamma + 2\partial_4 f + 2A = 0.$

We now make the substitution:

$$\begin{pmatrix} u \\ v \\ w \end{pmatrix} = \begin{pmatrix} 2x_1\left(1+\frac{x_1^2}{2}\right) & 2\left(1+\frac{x_1^2}{2}\right)^2 & x_1^2 \\ 2x_1 & x_1^2 & 2 \\ 1 & x_1 & 0 \end{pmatrix} \begin{pmatrix} \beta \\ \gamma \\ f \end{pmatrix} \Rightarrow$$

$$\begin{pmatrix} \beta \\ \gamma \\ f \end{pmatrix} = \frac{1}{4}\begin{pmatrix} -2x_1 & x_1^3 & 4+4x_1^2 \\ 2 & -x_1^2 & -4x_1 \\ x_1^2 & 2-\frac{x_1^4}{2} & -4x_1-2x_1^3 \end{pmatrix} \begin{pmatrix} u \\ v \\ w \end{pmatrix}$$

This leads to the revised set of equations:

$$\begin{cases}
\text{(i)} & \partial_1\alpha - 1 + A = 0 \\
\text{(ii)} & 2\partial_2\alpha + x_1\partial_1 v + 2\partial_1 w - u + \frac{1}{2}x_1^2 v = 0 \\
\text{(iii)} & 2\partial_3\alpha + \partial_1 u - x_1 u - x_1\left(1-\frac{x_1^2}{2}\right)v - (2-x_1^2)w = 0 \\
\text{(iv)} & 2\partial_4\alpha + \partial_1 v - 2w = 0 \\
\text{(v)} & x_1\partial_2 v + 2\partial_2 w + 2x_1\alpha + x_1^2 + 2(1+x_1^2)A = 0 \\
\text{(vi)} & x_1\partial_3 v + 2\partial_3 w + \partial_2 u + x_1^3 + (2+3x_1^2)\alpha + 2x_1(2+x_1^2)A = 0 \\
\text{(vii)} & x_1\partial_4 v + \partial_2 v + 2\partial_4 w + 2\alpha + 2x_1 + 4Ax_1 = 0 \\
\text{(viii)} & \partial_3 u + x_1(2+x_1^2)\alpha - 1 + \frac{x_1^4}{4} + 2\left(1+\frac{x_1^2}{2}\right)^2 A = 0 \\
\text{(ix)} & \partial_4 u + \partial_3 v + 2x_1\alpha + x_1^2 + 2Ax_1^2 = 0 \\
\text{(x)} & \partial_4 v + 1 + 2A = 0
\end{cases}$$

Then (i), (ii), (iv) and (x) give

$$\alpha = (1-A)x_1 + p$$

$$v = -(1+2A)x_4 + q$$

$$w = \partial_4\alpha + \frac{1}{2}\partial_1 v = \partial_4 p + \frac{1}{2}\partial_1 q$$

$$u = 2\partial_2\alpha + x_1\partial_1 v + 2\partial_1 w + \frac{1}{2}x_1^2 v$$

$$= 2\partial_2 p + x_1\partial_1 q + \partial_{11}q - \frac{1}{2}x_1^2 x_4(1+2A) + \frac{x_1^2}{2}q,$$

where $p = p(x_2, x_3, x_4)$ and $q = q(x_1, x_2, x_3)$. The remaining equations now impose the following conditions on p and q:

(a) $\partial_{111}q - 2x_1\partial_2 p + 2\partial_3 p - (2-x_1^2)\partial_4 p = 0$

(b) $2\partial_{24}p + \partial_{12}q + x_1\partial_2 q + 2x_1 p + 3x_1^2 + 2A = 0$

(c) $\partial_{112}q + x_1\partial_{12}q + \partial_{13}q + 2\partial_{22}p + 2\partial_{34}p + \frac{x_1^2}{2}\partial_2 q$
    $\quad + x_1\partial_3 q + (2+3x_1^2)p + 2x_1 + 4x_1^3 + (2x_1 - x_1^3)A = 0$

(d) $2\partial_{44}p + \partial_2 q + 2p + 3x_1 = 0$

(e) $\partial_{113}q + 2\partial_{23}p + x_1\partial_{13}q + \frac{x_1^2}{2}\partial_3 q + x_1(2+x_1^2)p - 1$
    $\quad + 2x_1^2 + \frac{5x_1^2}{4} + \left(2-\frac{x_1^4}{2}\right)A = 0$

(f) $2\partial_{24}p + \partial_3 q + 2x_1 p + \frac{5}{2}x_1^2 - x_1^2 A = 0.$

We obtain the following consequences:

(f) : $\quad \partial_3 q + 2x_1 p + \frac{5}{2}x_1^2 - x_1^2 A = -2\partial_{24} p$

$\partial_1(d)$: $\quad \partial_{12} q + 3 = 0 \quad (\Rightarrow \partial_{112} q = 0)$

$\partial_1(b)$: $\quad \partial_{112} q + x_1 \partial_{12} q + \partial_2 q + 2p + 6x_1 = 0 \Rightarrow \partial_2 q + 2p + 3x_1 = 0$

(d) : $\quad \partial_{44} p = 0 \Rightarrow p = x_4 s + t$,

where $s = s(x_2, x_3)$ and $t = t(x_2, x_3)$. Then

$$(b): \quad 2\partial_2 s = 3 - 2A \Rightarrow s = \left(\frac{3 - 2A}{2}\right) x_3 + \xi,$$

where $\xi = \xi(x_3)$. Also

$$\partial_{24} p = \partial_2 s = \frac{3 - 2A}{2} \Rightarrow \partial_3 q = 2A - 3 - 2x_1 p - \frac{5}{2}x_1^2 + x_1^2 A$$

$$\Rightarrow \partial_{13} q = -2p - 5x_1 + 2x_1 A \to \partial_{113} q = -5 + 2A.$$

Then (e) is equivalent to

$$2\partial_{23} t - 6 - \frac{9x_1^2}{2} + 4A + 3x_1^2 A = 0.$$

Taking ∂_1 of this then yields $A = 3/2$ so that $s = \xi(x_3)$ and $\partial_{23} t = 0$. We therefore have

$$\partial_2 t = \sigma(x_2), \qquad p = x_4 \xi(x_3) + t(x_2, x_3).$$

for a function $\sigma = \sigma(x_2)$. Equation (c) now shows that $\sigma' = -\xi' = a$ constant, so that

$$\sigma = ax_2 + b, \quad \xi = -ax_3 + c, \quad \partial_2 t = ax_2 + b \Rightarrow t = \frac{ax_2^2}{2} + bx_2 + e(x_3),$$

for constants a, b, c and for some function $e = e(x_3)$. We now have

$$p = x_4(-ax_3 + c) + \frac{ax_2^2}{2} + bx_2 + e(x_3)$$

$$\partial_3 q = 2A - 3 - 2x_1 p - \frac{5}{2}x_1^2 + x_1^2 A$$

$$\partial_2 q = -2p - 3x_1.$$

Then $\partial_{23} q = \partial_{32} q \Rightarrow x_1 \partial_2 p = \partial_3 p \Leftrightarrow x_1(ax_2 + b) = -ax_4 + e'(x_3) \Leftrightarrow a = b = 0$ and $e' = $ const. Thus

$$p = cx_4 + e \quad \text{and} \quad \partial_2 q = -2p - 3x_1 = -2cx_4 - 2e - 3x_1.$$

But since q is independent of x_4, we must have $c = 0$ so that $p = e$. Also

$$\partial_2 q = -3x_1 - 2e \Rightarrow q = -3x_1 x_2 - 2ex_2 + \rho(x_1, x_3),$$

for some function $\rho = \rho(x_1, x_3)$, so that

$$\partial_3 q = \partial_3 \rho = -2ex_1 - x_1^2 \Rightarrow \rho = -2ex_1x_3 - x_1^2x_3 + r(x_1),$$

for some function $r = r(x_1)$. But now

$$(a): \quad \partial_{111}q = 0 \Rightarrow r''' = 0 \Rightarrow r = \frac{ax_1^2}{2} + bx_1 + c,$$

for some (new) constants a, b, c. We conclude that

$$
\begin{aligned}
p &= e \\
q &= -3x_1x_2 - x_1^2x_3 - 2ex_2 - 2ex_1x_3 + \tfrac{ax_1^2}{2} + bx_1 + c \\
\alpha &= -\tfrac{x_1}{2} + e \\
v &= -4x_4 - 3x_1x_2 - x_1^2x_3 - 2ex_2 - 2ex_1x_3 + \tfrac{ax_1^2}{2} + bx_1 + c \\
w &= \tfrac{1}{2}(-3x_2 - 2x_1x_3 - 2ex_3 + ax_1 + b) \\
u &= x_1(-3x_2 - 2x_1x_3 - 2ex_3 + ax_1 + b) - 2x_3 + a - 2x_1^2x_4 \\
&\quad + \tfrac{x_1^2}{2}(-3x_1x_2 - x_1^2x_3 - 2ex_2 - 2ex_1x_3 + \tfrac{ax_1^2}{2} + bx_1 + c),
\end{aligned}
$$

for constants a, b, c, e, which gives

$$\beta = -\frac{3x_2}{2} + a\frac{x_1}{2} - 2ex_3 + b$$

$$\gamma = -x_3 + \frac{a}{2}$$

$$f = -2x_4 - ex_2.$$

It is easily seen that the arbitrary constants correspond to the addition of a Killing vector field to E, so we can summarize the above calculations in the following theorem.

Theorem 3.5.1 *The geometry* Nil_4 *admits a unique soliton structure, with flow defined up to addition of a Killing vector field, given by*

$$E = -\frac{x_1}{2}\partial_1 - \frac{3x_2}{2}\partial_2 - x_3\partial_3 - 2x_4\partial_4.$$

This result is confirmed by Theorem 3.4.2, where we take $\varphi : \mathrm{Nil}_4 \to \mathrm{Nil}_3$ to be the natural projection and exploit the soliton structure on Nil_3 given by Example 3.3.3; indeed, we apply Theorem 3.4.2 with $\lambda \equiv 1$, $\theta = dx_4 + x_1dx_2 + \tfrac{x_1^2}{2}dx_3$ and $X = -\tfrac{x_1}{2}\partial_1 - \tfrac{3x_2}{2}\partial_2 - x_3\partial_3$. However, 3.4.2 does not establish uniqueness.

References

[1] P. Baird, *A class of three-dimensional Ricci solitons*, Geometry and Topology **13** (2009), 979–1015.

[2] P. Baird and L. Danielo, *Three-dimensional Ricci solitons which project to surfaces*, J. reine angew. Math., **608** (2007), 65–91.

[3] P. Baird and J. C. Wood, Harmonic Morphisms between Riemannian Manifolds, London Math. Soc. Monograph (New Series), vol. 29, Oxford University Press, 2003.

[4] R. L. Bryant, *Ricci flow solitons in dimension three with* SO(3)-*symmetries*, preprint, Duke Univ., Jan 2005.

[5] B. Chow, S-C. Chu, D. Glickenstein, C. Guenther, J. Isenberg, T. Ivey, D. Knopf, P. Lu, F. Luo and L. Ni, The Ricci flow: Techniques and Applications, Part 1: Geometric aspects, AMS Mathematical Surveys and monographs, **135**, 2007.

[6] B. Chow and D. Knopf, The Ricci flow: An Introduction, Mathematical Surveys and Monographs, Vol. 110, American Mathematical Society, Providence, RI, 2004.

[7] C. Guenther, J. Isenberg and D. Knopf, *Stability of the Ricci flow at Ricci-flat metrics*, Comm. Anal. Geom. **10** (2002), no. 4, 741–777.

[8] C. Guenther, J. Isenberg and D. Knopf, *Stability of Ricci nilsolitons*, preprint (2006).

[9] R. Hamilton, *A compactness property for solutions of the Ricci flow*, Amer. J. Math. **117** (1995), 545–572.

[10] T. Ivey, *Ricci solitons on compact three-manifolds*, Differential Geom Appl. **3** (4) (1993), 301–307.

[11] T. Ivey, *New examples of complete Ricci solitons*, Proc. Amer. Math. Soc. **122** (1994), 241–245.

[12] J. Lauret, *Ricci soliton homogeneous nilmanifolds*, Math. Ann. **319** (2001), 715–733.

[13] J. Lott, *On the long-time behaviour of type-III Ricci flow solutions*, Math. Annalen, **339**, No. 3 (2007), 627–666.

[14] G. Perelman, *The entropy formula for the Ricci flow and its geometric applications*, arXiv:math.DG/0211159.

[15] N. Sesum, *Linear and dynamical stability of Ricci flat metrics*, Duke. Math. J., **133** (2006), 1–26.

Author's address:

Département de Mathématiques,
Université de Bretagne Occidentale,
6 Avenue Le Gorgeu, – CS 93837
29238 Brest,
France
Paul.Baird@univ-brest.fr

4

Open Iwasawa cells and applications to surface theory

JOSEF F. DORFMEISTER

4.1 Introduction

In recent years, many surfaces of a special type, like surfaces of constant mean curvature (CMC) in $\mathbb{R}^3$, Willmore surfaces in S^n or spacelike mean curvature surfaces in Minkowski space $\mathbb{L}^3$, have been constructed using loop groups. The procedure in all these cases is fairly similar: one considers a 'Gauss type map' from a Riemann surface M to some real symmetric space G/K, like the classical Gauss map or a conformal Gauss map, and characterizes a class of surfaces by the harmonicity (or the conformal harmonicity) of this Gauss map. Lifting the Gauss map to a map F from the universal cover $\tilde{M}$ into G we obtain a 'moving frame'. It has been shown in [7] how all such frames can be constructed from holomorphic data: considering the moving frame for each member of the associated family one obtains an 'extended frame' F_λ and one can show that it suffices to construct all such extended frames, since, as an added feature, for all these special surface classes there exists a simple formula, 'Sym type formula', which reconstructs the given immersion from its extended frame.

The heart of the loop group method for the construction of surfaces thus is a procedure that produces all extended frames of all surfaces of the special classes to which this method applies. This procedure involves loop group decompositions. For simplicity, let's consider conformal immersions like CMC and spacelike CMC from some simply connected domain $\mathbb{D} \subset \mathbb{C}$ into $\mathbb{R}^3$ and $\mathbb{L}^3$ respectively. Then the extended frame F has, for all $z \in \mathbb{D} \setminus S$, S a discrete subset of $\mathbb{D}$, a 'Birkhoff decomposition' $F = F_- L_+$. (Here F_- only contains non-positive powers of λ and starts with I and L_+ only contains non-negative powers of λ in a Fourier expansion.) The matrix function F_- maps into $G^{\mathbb{C}}$ and is meromorphic with poles at the points of S. The Maurer-Cartan form $\eta = F_-^{-1} dF_-$ of F_- is a meromorphic $(1, 0)-$ form and is called the 'normalized potential' for the given immersion. It is of the form $\eta = \lambda^{-1}\eta_{-1}(z)dz$ and has values in $p^{\mathbb{C}}$, where $g = k + p$ is the Cartan decomposition of the symmetric space G/K. As a consequence, if one wants to construct an extended frame for a surface in one of the classes under consideration, one will start with some normalized potential $\eta = \lambda^{-1}\eta_{-1}(z)dz$, a meromorphic $(1, 0)$−form with values in $p^{\mathbb{C}}$, which has a meromorphic solution to the ODE $dF_- = F_-\eta$. In the next step one decomposes F_- in the form $F_- = FW_+$, with $F \in G$. If G is maximal compact in the complex Lie group $G^{\mathbb{C}}$, then this decomposition always exists and is called 'Iwasawa decomposition'. But in many cases relevant to surface theory G is non-compact. In the finite dimensional case such a situation has been investigated by Aomoto, Matsuki and Rossman [1], [13], [15]. In the (infinite dimensional) loop group case similar results have been obtained by Kellersch [10], [11]. It turns out that a decomposition of the form $F_- = FW_+$ only exists away from a singular set $\mathbb{S}$. This singular set may consist of points, but may also contain curves. The matrix function F is, where defined, the extended frame of a harmonic map associated with surfaces of our class. Then a 'Sym formula' produces the actual surface. More precisely, the Sym formula produces a map, called here 'weak immersion', which is an immersion of the desired surface type, wherever its differential has rank two.

In the construction outlined above, three types of singularities occur: firstly, poles in the normalized potential. Secondly, singularities due to the non-global Iwasawa splitting and, thirdly, branch points or branching curves, where the differential of the mapping defined by the Sym formula drops rank.

The first kind of singularities can be taken care of by considering a slightly different construction, producing holomorphic potentials (with a less trivial Fourier expansion). The last kind of singularities is sometimes unavoidable and will not be discussed in this note.

Singularities of the second kind, stemming from the non-globality of the Iwasawa splitting for non-compact G, have been investigated by a few authors only so far and for $G = Sl(2, \mathbb{R})$ only, to the best of our knowledge. It turns out [2] that the coefficients of the frame F have strong singularities along the singular set $\mathbb{S}$, while the actual weak immersion may be smooth across $\mathbb{S}$ with a differential of rank lower than two.

To understand this behaviour better, we consider general symmetric spaces G/K and the corresponding Iwasawa decompositions $LG_\sigma^{\mathbb{C}} = \bigcup_{\delta \in \Xi} LG_\sigma \cdot \delta \cdot L^+G_\sigma^{\mathbb{C}}$, where Ξ is a set of representatives for the obvious action of $LG_\sigma \times L^+G_\sigma^{\mathbb{C}}$ on $LG_\sigma^{\mathbb{C}}$. These cosets will be called 'Iwasawa cells'. Similar to what was found in the $Sl(2, \mathbb{R})$ case one also finds, for general real G, that in general several open Iwasawa cells exist. Clearly, considering F_- as above, we obtain a well-behaved extended frame of some harmonic map as long as F_- stays in an open Iwasawa cell. We are therefore interested in a description of the open Iwasawa cells and their boundary.

As main result of this paper we prove a first step in this direction:

Theorem 4.1.1 *(1) The union of all open Iwasawa cells is dense in $LG_\sigma^{\mathbb{C}}$.*

(2) There exists a line bundle $\mathcal{L}^*$ over $LG_\sigma^{\mathbb{C}}$ and a real analytic section α^* of $\mathcal{L}^*$ such that some $g \in LG_\sigma^{\mathbb{C}}$ is in an open Iwasawa cell only if $\alpha^*(g) \neq 0$.

In applications to surface theory we will find real analytic, complex valued functions (not only sections of some bundle) such that these functions do not vanish if $F_-(z, \lambda)$ is in an open Iwasawa cell.

4.2 Basic notation and the Birkhoff decomposition

Let G be a real form of the simply connected, complex, semi-simple, matrix Lie group $G^{\mathbb{C}}$. Then we consider the group $LG^{\mathbb{C}}$ of loops in $G^{\mathbb{C}}$, i.e. the set of all maps from the unit circle S^1 into G^C, where we assume that each matrix entry has a Fourier expansion for which the sum of absolute values of its coefficients, multiplied with a weight, converges (weighted Wiener topology). For more details we refer to [4].

Let τ denote the anti-holomorphic involution of $G^{\mathbb{C}}$ with $G = Fix(\tau)$.

In this paper we consider exclusively inner symmetric spaces G/K. We can assume that there exists a holomorphic involution σ such that $K = Fix(\sigma)$ and that σ and τ commute. For a description of harmonic maps into symmetric spaces it is advantageous to use 'twisted loop groups'.

The following groups will be used throughout this paper.

1 $LG_\sigma^{\mathbb{C}} = \{g \in LG^{\mathbb{C}}; \sigma(g(\lambda)) = g(-\lambda)\}$,

2 $L^+G_\sigma^{\mathbb{C}} = \{g \in LG_\sigma^{\mathbb{C}}; g$ has a holomorphic extension to the open unit disk$\}$,

3 $L_*^+G_\sigma^{\mathbb{C}} = \{g \in LG_\sigma^{\mathbb{C}}; g$ has a holomorphic extension to the open unit disk and $g(0) = I\}$,

4 $L^-G_\sigma^{\mathbb{C}} = \{g \in LG_\sigma^{\mathbb{C}}; g$ has a holomorphic extension to the upper hemisphere$\}$,

5 $L_*^-G_\sigma^{\mathbb{C}} = \{g \in LG_\sigma^{\mathbb{C}}; g$ has a holomorphic extension to the upper hemisphere and $g(\infty) = I\}$,

6 $LG_\sigma = \{g \in LG^{\mathbb{C}}; g(\lambda) \in G$ for all $\lambda \in S^1\}$.

Note, we identify the unit disk with the lower hemisphere of the Riemann sphere S^2. And holomorphicity in the upper hemisphere means that after changing $\lambda \to \lambda^{-1}$ one has a holomorphic quantity on the open unit disk.

Using this notation we can state the first important decomposition theorem (see e.g. [14],[3]).

Theorem 4.2.1 (Birkhoff decomposition Theorem)

$$LG_\sigma^{\mathbb{C}} = \bigcup_{\omega \in \Omega} L^- G_\sigma^{\mathbb{C}} \cdot \omega \cdot L^+ G_\sigma^{\mathbb{C}} \qquad (4.1)$$

In particular, every $g \in LG_\sigma^{\mathbb{C}}$ can be written in the form

$$g = g_- \omega g_+ \qquad (4.2)$$

for some $g_* \in L^* G_\sigma^{\mathbb{C}}$ and $\omega \in \Omega$. The set Ω for the disjoint union above is closely related to the Weyl group of $LG_\sigma^{\mathbb{C}}$ (see e.g. [14]).

Moreover, the group multiplication $L^- G_\sigma^{\mathbb{C}} \times L^+ G_\sigma^{\mathbb{C}} \to LG_\sigma^{\mathbb{C}}$ is an analytic diffeomorphism onto an open dense subset of $LG_\sigma^{\mathbb{C}}$.

Remark 4.2.2 (1) The Birkhoff decomposition for g in the 'big (left Birkhoff) cell'

$L^- G_\sigma^{\mathbb{C}} \cdot L^+ G_\sigma^{\mathbb{C}}$ can be made unique if one requires $g_- \in L_*^- G_\sigma^{\mathbb{C}}$ or $g_+ \in L_*^+ G_\sigma^{\mathbb{C}}$.

(2) For the other cells one can also formulate a condition that makes the representation unique [9],[14].

4.3 Iwasawa decomposition

Let $G^{\mathbb{C}}$, τ and G be as before. Then we consider the group action

$$(LG_\sigma \times L^+ G_\sigma^{\mathbb{C}}) \times LG_\sigma^{\mathbb{C}} \to LG_\sigma^{\mathbb{C}}, \quad \text{where} \quad (g, v_+).h = ghv_+^{-1} \qquad (4.3)$$

Let Ξ denote a set of representatives for the orbits (= double cosets) of this group action. Then we obtain a general form (i.e. without any precise description of Ξ) of the "Iwasawa decomposition".

Theorem 4.3.1 (Iwasawa Decomposition)

$$LG_\sigma^{\mathbb{C}} = \bigcup_{\delta \in \Xi} LG_\sigma \cdot \delta \cdot L^+ G_\sigma^{\mathbb{C}} \qquad (4.4)$$

In particular, every $g \in LG_\sigma^{\mathbb{C}}$ can be written in the form

$$g = h \delta g_+ \qquad (4.5)$$

for some $h \in LG_\sigma$, $g_+ \in L^+ G_\sigma^{\mathbb{C}}$ and $\delta \in \Xi$.

Remark 4.3.2 (1) The finite dimensional case of an Iwasawa decomposition for any real form G of some complex semisimple Lie group $G^{\mathbb{C}}$ has been investigated by Aomoto [1]. More general situations have been considered by Matsuki [13] and Rossman [15]

(2) The case of untwisted loop groups has been investigated by Kellersch [10], [11].

4.4 Iwasawa decomposition via Birkhoff decomposition

Assume that G is maximal compact in $G^{\mathbb{C}}$, then the Iwasawa Decomposition in $LG^{\mathbb{C}}$ is global, i.e. for every $g \in LG^{\mathbb{C}}$ there exist $h \in LG$ and $v_+ \in L^+G^{\mathbb{C}}$ such that $g = hv_+$. This is equivalent to that for every $g \in LG^{\mathbb{C}}$ there exists some $v_+ \in L^+G^{\mathbb{C}}_\sigma$ such that $\tau(g)^{-1}g = \tau(v_+)^{-1}v_+$ holds. In particular, for every $g \in LG^{\mathbb{C}}$ the matrix $\tau(g)^{-1}g$ is in the big Birkhoff cell.

What we just explained works equally well in the case of twisted loop groups.

The trick to consider a Birkhoff decomposition of $\tau(g)^{-1}g$ also works (to some extent–see below) in the case of arbitrary real forms G of $G^{\mathbb{C}}$. However, in the general case it is not possible to prove that $\tau(g)^{-1}g$ is in the big Birkhoff cell for every $g \in LG^{\mathbb{C}}_\sigma$. And even if $\tau(g)^{-1}g$ is in the big Birkhoff cell, it does not follow that $\tau(g)^{-1}g = \tau(v_+)^{-1}v_+$ holds for some $v_+ \in L^+G^{\mathbb{C}}_\sigma$.

Below we will discuss this situation for twisted loop groups. The case of an untwisted loop group is very similar. Since it has been investigated in much detail by Kellersch [10],[11] we will not address it in this note.

Starting from the expression $\tau(g)^{-1}g$ one tries to simplify it by transformations of the type $\tau(g)^{-1}g \rightarrow \tau(v_+)(\tau(g)^{-1}g)v_+^{-1}$ as much as possible (we will explain below how we interpret this statement). Once one has obtained this way an expression, say β, that cannot be simplified any more, one writes $\beta = \tau(\delta)^{-1}\delta$ (what can be done by construction) and altogether obtains $gv_+^{-1}\delta^{-1} = \tau(gv_+^{-1}\delta^{-1})$. Calling this expression h one obtains the Iwasawa decomposition $g = h\delta v_+$ of g.

In view of the equality $\tau(v_+) \cdot (\tau(g)^{-1}g) \cdot v_+^{-1} = \beta$ we consider decompositions of the form $\tau(g)^{-1}g = \tau(v_+)^{-1} \cdot \beta \cdot v_+$. This reminds one of a Birkhoff decomposition. However, since we require here that the element in $L^-G^{\mathbb{C}}_\sigma$ is related to the element in $L^+G^{\mathbb{C}}_\sigma$, it is, a priori, not clear, that β can be chosen as a representative of a Weyl group element.

Theorem 4.4.1 ([10],[11], Theorem 3.64) *Let $G^{\mathbb{C}}$ be a simply-connected, semisimple, complex matrix Lie group and τ an anti-holomorphic involution with fixed point group G. Assume that G/K is an inner symmetric space, where*

$K = Fix(\sigma)$ *for some inner involution* σ *of* G. *Let* U *be a maximal compact subgroup of* $G^{\mathbb{C}}$ *which is invariant under* σ *and* τ *and* T *a maximal torus in* U *which is fixed pointwise by* σ. *Let* N *denote the group of Laurent polynomials in* $LG_{\sigma}^{\mathbb{C}}$ *which normalize* T.

Then for every $g \in LG_{\sigma}^{\mathbb{C}}$ *there exists some* $v_+ \in L^+G_{\sigma}^{\mathbb{C}}$ *and some* $n \in N$ *such that*

$$\tau(g)^{-1}g = \tau(v_+)^{-1} \cdot n \cdot v_+. \qquad (4.6)$$

In particular, we have $n \in LG_{\sigma}^{\mathbb{C}}$.

In [10],[11] and in the untwisted case the corresponding result is Proposition 3.36 and the choice of n is specified much further. In the twisted case no similar detailed description of the 'middle term' Ξ has been given yet in a general setting. However, we will make a first step towards a detailed description of the open Iwasawa cells.

For this we note that Lemma 4.37, Proposition 4.38 and Proposition 4.39 of [10],[11] still hold in the twisted case.

Theorem 4.4.2 *With the notation of the last theorem we can assume w.l.g. that* n *has the form* $n = qt$, *where* q *is in the normalizer of* T *in* U *and* t *is a homomorphism from* S^1 *into* T. *In particular,* n *is a representative of a Weyl group element of the twisted loop group* $LG_{\sigma}^{\mathbb{C}}$.

Proof Consider the automorphism $\hat{\sigma}$ of $LG^{\mathbb{C}}$ given by $\hat{\sigma}(g)(\lambda) = \sigma(g(-\lambda))$. Clearly, $LG_{\sigma}^{\mathbb{C}} = Fix(\hat{\sigma})$. Kellersch's result above, Theorem 4.1, simply states that the stated decompositions are valid with all occurring matrices fixed by $\hat{\sigma}$.

It thus suffices to verify that the three results quoted above and proven in [10],[11] for the untwisted case still hold with all occurring matrices fixed under $\hat{\sigma}$. For the generalization of loc.cit Lemma 4.37 one observes that $n = \hat{\sigma}(n) = \hat{\sigma}(q) \cdot \hat{\sigma}(exp(H_-)) \cdot \hat{\sigma}(t)$ and that $\hat{\sigma}$ maps q, $exp(H_-)$ and t onto matrices with the same properties as stated in loc.cit. Therefore, by the uniqueness result for the representation (in the untwisted case), we obtain that q, $exp(H_-)$ and t are fixed under $\hat{\sigma}$. The equations of loc.cit. Proposition 4.38 are valid anyway for q, $exp(H_-)$ and t. And the same statement holds for the proof of loc.cit Proposition 4.39. $\qquad \square$

Corollary 4.4.3 An element $g \in LG_{\sigma}^{\mathbb{C}}$ is in the open Iwasawa cell $LG_{\sigma} \cdot L^+G_{\sigma}^{\mathbb{C}}$ if and only if $t(\lambda) = I$ and $q = \tau(b)^{-1}b$ for some $b \in K^{\mathbb{C}}$.

Proof Note that $g = hv_+$ is equivalent with $\tau(g)^{-1}g = \tau(v_+)^{-1}v_+$. The realization $\tau(g)^{-1}g = \tau(v_+)^{-1} \cdot qt \cdot v_+$ shows that g is as required only if $\tau(g)^{-1}g$ is in the big Birkhoff cell. This happens only if $qt(\lambda) \in K^{\mathbb{C}}$. Since this matrix is in particular independent of λ we obtain $t(\lambda) \equiv I$ and $q \in K^{\mathbb{C}} \cap U$. Now the claim follows.   □

4.5 A function defining the open Iwasawa cells

The main goal of this section is to prove that there exists a real analytic section of some line bundle $\mathcal{L}^*$ over $LG_\sigma^{\mathbb{C}}$ which contains the union of all open Iwasawa cells in its non-vanishing set. In geometric applications we usually use loop matrices which are real analytic on S^1. But in our general setting we have only considered loop matrices in some Banach algebra defined by some weighted Wiener norm. In particular, until now the Wiener norm, given by the sum of all absolute values of all Fourier coefficients, worked for us.

For the result below, however, we need a weighted Wiener norm, where the weight is of the form $w(r) = (1 + |nr|)^m$ with $0 < m, n$.

Theorem 4.5.1 *There exists a real analytic line bundle $\mathcal{L}^*$ over $LG_\sigma^{\mathbb{C}}$ and a real analytic section α^* of this bundle such that some $g \in LG_\sigma^{\mathbb{C}}$ is in an open Iwasawa cell only if $\alpha^*(g) \neq 0$.*

Proof In this proof we use substantially ideas and results from [16]. First we consider the Grassmannian $Gr^{(n)}$ and the group $Aut(Gr^{(n)}))$ of holomorphic automorphisms of $Gr^{(n)}$, where n is given by the realization $G^{\mathbb{C}} \subset Gl(n, \mathbb{C})$. Let $p : LG^{\mathbb{C}} \to Aut(Gr^{(n)}))$ denote the map associating with a loop matrix $A(\lambda) = \sum_{\mathbb{Z}} A_j \lambda^j$ the infinite matrix which has in the j−th parallel to the diagonal the string of matrices which are all equal to A_j. Note that this is a linear mapping. Moreover, using a weighted Wiener norm on the loop groups and an appropriate norm on $Gr^{(n)}$ and on $Aut(Gr^{(n)}))$, this linear map induces a holomorphic map ([4], section 3) from $LG_\sigma^{\mathbb{C}}$ to $Aut(Gr^{(n)}))$. As a consequence, the map $A(\lambda) \to p(\tau(A)^{-1}A))$ is real analytic. Applying the image of this map to the base point H_+ of $Gr^{(n)}$ we obtain a real analytic map $f(A) = p(\tau(A)^{-1}A)).H_+$ from $LG_\sigma^{\mathbb{C}}$ to $Gr^{(n)}$, since the automorphism group of $Gr^{(n)}$ acts holomorphically. Next we use the holomorphic section α of the Det^* bundle over $Gr^{(n)}$ introduced in [16]. Pulling back the Det^* bundle along f we obtain the real analytic bundle $\mathcal{L}^* = f^*Det^*$ over $LG_\sigma^{\mathbb{C}}$ and a real analytic section α^* of this pull back bundle. We know (see e.g. [7]) that α^* has a non-zero value exactly at all $A(\lambda)$ for which $\tau(A)^{-1}A$ is in the big Birkhoff cell.

Now assume that g is in an open Iwasawa cell $LG_\sigma \cdot \delta \cdot L^+G^{\mathbb{C}}_\sigma$. Then not only for the given g, but for all b in this open cell, we obtain $\tau(b)^{-1}b = \tau(u_+)^{-1} \cdot \omega \cdot u_+$ for some $u_+ \in L^+G_\sigma^{\mathbb{C}}$ and $\omega = \tau(\delta)^{-1}\delta$. Assume one such element is not in the big Birkhoff cell. Then we can assume $\omega = qt$ with non-trivial homomorphism t from S^1 to T. Therefore all these elements are not in the big Birkhoff cell and therefore α^* vanishes for each such b. But then the real analytic section α^* vanishes on an open subset of $LG_\sigma^{\mathbb{C}}$. Since σ is inner, the twisted loop group $LG_\sigma^{\mathbb{C}}$ is isomorphic with the untwisted loop group $LG^{\mathbb{C}}$. And since we have assumed that $G^{\mathbb{C}}$ is simply-connected, the untwisted loop group is connected. As a consequence, α^* vanishes on the full group. This is a contradiction, since α^* does not vanish at $I \in LG_\sigma^{\mathbb{C}}$. Therefore $t \equiv I$ and $\omega = q \in K^{\mathbb{C}} \cap U$ follows. $\qquad\square$

Proposition 4.5.2 *With the notation of the preceding Theorem, assume* $\alpha^*(g) \neq 0$. *Then* $\tau(\delta)^{-1}\delta = q \in K^{\mathbb{C}} \cap U$. *Moreover, if in addition* $Lg_\sigma^{\mathbb{C}} = Lg_\sigma + \delta^{-1}L^+ g_\sigma^{\mathbb{C}}\delta$, *then* δ *defines an open Iwasawa cell.*

Proof Consider the infinitesimal stabilizer of the action of $LG_\sigma \times L^+ G_\sigma^{\mathbb{C}}$ on δ. We obtain the equation $h\delta - \delta w_+ = 0$. This implies $h = \delta^{-1}w_+\delta$, and, since both sides are fixed by τ, a simple calculation shows that this equation is equivalent with $\tau(w_+) = q^{-1}w_+ q$. As a consequence, w_+ does not contain any λ and thus $w_+ = w_0 \in K^{\mathbb{C}}$. But then $w_0 = x + iy$ with $x, y \in Lie(K)$ and $\tau(w_0) = q^{-1}w_0 q$ is equivalent with $q^{-1}xq = x$ and $q^{-1}yq = -y$. Since $Ad(q)$ only has the eigenvalues ± 1, it follows that the stabilizer of the group action on δ has the dimension of K. This is clearly also the dimension of the action of the same group on I. For the latter case it is easy to see that the orbit is open in $LG_\sigma^{\mathbb{C}}$.

In view of our assumption we can conclude from the equality of the dimensions of the stabilizers that also the orbit through δ is open in $LG_\sigma^{\mathbb{C}}$. $\qquad\square$

The proofs just given show moreover:

Corollary 4.5.3 (1) An Iwasawa cell $LG_\sigma \cdot \delta \cdot L^+ G_\sigma^{\mathbb{C}}$ is open in $LG_\sigma^{\mathbb{C}}$ only if the element $\tau(\delta)^{-1}\delta$ is in the big Birkhoff cell. Under the additional assumption in the last proposition, the converse also holds.

(2) If q is as above, then $\{\tau(w)^{-1} \cdot q \cdot w; w \in K^{\mathbb{C}}\}$ always has dimension $\dim K$.

Even if the union of all open Iwasawa cells is not equal to the non-vanishing set of the real analytic section discussed above, we still obtain

Proposition 4.5.4 *The union of all open Iwasawa cells is dense in* $LG_\sigma^{\mathbb{C}}$.

Proof If the union of all open Iwasawa cells is not dense, then there exists an open set in $G_\sigma^{\mathbb{C}}$ of the complement of this union such that the section α^* does not vanish anywhere on this open set. To each such point there corresponds some q as in Proposition 4.2.

We note that every such q can be written in the form $q = a\rho$, where $a \in T$ and ρ is a once and for all fixed representative of the Weyl group of $K^{\mathbb{C}}$ relative to T. Thus freedom can only come from the fact that some $q = a\rho$, but also some $\hat{q} = \hat{a}\rho$ occurs. If the claim of the proposition is wrong, then there exists some q to which uncountably many $\hat{q}$'s belong. But then one of the a's would be a cluster point. Now (2) in the corollary above shows that this is not possible for q's representing different Iwasawa cells.   □

Remark 4.5.5 (1) In the finite dimensional case, Aomoto [1] has given a very satisfactory description of all occurring cells. A fairly complete description of all Iwasawa cells has been given in the untwisted case by Kellersch [10],[11]. For the twisted case much remains to be done.

4.6 Applications to surface theory

For the general procedure we refer to the introduction. We will list a few concrete examples.

Example 1: We consider space like surfaces in Minkowski space $\mathbb{L}^3$ with inner product $\langle x, y \rangle = -x_0 y_0 + x_1 y_1 + x_2 y_2$. For each such immersion one also considers a 'Gauss map', the future pointing unit normal vector. One can show that the spacelike immersion has constant mean curvature (CMC) if and only if the Gauss map is harmonic. If the immersion is defined on some simply connected domain $\mathbb{D}$, we thus need to construct all harmonic maps from $\mathbb{D}$ to $\mathbb{H}^2 = \{v \in \mathbb{L}; \langle v, v \rangle = -1, v_0 > 0\}$. Since $\mathbb{H}^2 \cong SU(1, 1)/K$, where $K \cong U(1)$, is realized by the diagonal matrices in $SU(1, 1)$, is a symmetric space, the loop group method is applicable.

The complex Lie group thus is $G^{\mathbb{C}} = Sl(2, C)$ with real form $G = SU(1, 1)$ and stabilizer group K for the symmetric space under consideration. The complex anti-linear involution τ is given by $\tau(g) = \sigma_3(\bar{g}^t)^{-1}\sigma_3$, where σ_3 denotes the diagonal matrix with entries 1 and -1. The involution σ, defining the symmetric space, is given by conjugation by σ_3.

As outlined already in the introduction, each spacelike CMC immersion can be constructed from some "potential" η. In our case, η can be chosen of the form $\eta = \lambda^{-1}\eta_{-1}$, where η_{-1} is off-diagonal and meromorphic and we can assume that the solution to the ODE $dC = C\eta$ is also meromorphic. For convenience we choose a base point z_* and assume $C(z_*, \lambda) = I$. Moreover, to make the presentation somewhat smoother we assume that C is holomorphic in $\mathbb{D}$.

Until now the setting does not only apply to spacelike CMC surfaces, but also to CMC surfaces in $\mathbb{R}^3$ and timelike constant negative Gauss curvature surfaces in $\mathbb{L}^3$. The distinction happens, when one applies the Iwasawa decomposition to C.

In the case of CMC surfaces in $\mathbb{R}^3$ the Iwasawa decomposition is global and thus does not introduce any new singularity. But in the case under consideration, the Iwasawa splitting is not global. As a matter of fact, fitting to the result in the last section we have two open Iwasawa cells, defined by $q = I$ and $q = -I$.

(The corresponding δ is $\delta = I$ in the first case and in the second case the off-diagonal matrix with $(1, 2)$ – entry λ and $(2, 1)$ – entry $-\lambda^{-1}$, see e.g. [2].)

Clearly, we expect problems, where C crosses from one open cell to the other one. We obtain in this example that the boundary between these two cells is given by the vanishing set of the section α^*. One can describe this here in more detail: We consider the map $\mathbb{D} \to LSl(2, C)_\sigma$, $z \to C(z, \lambda)$ and pull back the bundle $\mathcal{L}^*$. Since $\mathbb{D}$ is contractible, this pull back bundle is trivial and the section α^* corresponds to a real analytic function $\mathcal{F} : \mathbb{D} \to \mathbb{C}$. Hence the points in $\mathbb{D}$, where the matrix function $C(z, \lambda)$ touches the boundary between the open Iwasawa cells, are exactly the points where the function $\mathcal{F}$ vanishes.

Let's start now with C at z_*. Then $C = C(z, \lambda)$ starts at I. As long as z varies such that $C(z, \lambda)$ stays inside the same open Iwasawa cell as I we can decompose $C = FV_+$ with $F \in LSU(1, 1)_\sigma$ and $V_+ \in L^+Sl(2, C)_\sigma$.

In [2], the behaviour of F has been investigated, when z approaches the set $\mathcal{F} \equiv 0$. It was shown there that the coefficients of F behave singularly. However, at least in some cases, the actual immersion, obtained from F by a Sym type formula does not necessarily behave very singularly: in some cases it stays real analytic when crossing certain parts of the set $\mathcal{F} \equiv 0$, but its differential drops rank, so that it is no longer an immersion outside of the open Iwasawa cells.

Remark 4.6.1 In the case of quantum cohomology of $\mathbb{C}P^1$, naturally a harmonic map of a spacelike CMC surface in $\mathbb{L}^3$ occurs. The main interest in this context is to choose this surface such that its frame never leaves the open cell it starts in.

More generally, such a requirement is imposed in tt^*–geometry [5].

Example 2: In this example we consider Willmore surfaces in S^n. To keep the presentation short we will be quite sketchy here. Let's start with some conformal immersion $y : \mathbb{D} \to S^n$. Then one constructs canonically some "conformal Gauss map" $f : \mathbb{D} \to Gr_{1,3}(\mathbb{L}^{n+2})$, the Grassmannian of four-dimensional subspaces of $\mathbb{L}^{n+2}$ with induced Minkowski metric (thus of signature $(1, 3)$.

It is well known that the conformal Gauss map is conformally harmonic if and only if the original conformal immersion y is a Willmore surface. Thus we need to consider conformally harmonic maps from $\mathbb{D}$ into the symmetric space $Gr_{1,3}(\mathbb{L}^{n+2}) \cong SO(1, n + 1)/SO(1, 3) \times SO(n - 2)$. As a consequence, the loop group method is applicable.

In this case the complex Lie group $G^{\mathbb{C}}$ is $SO(n + 2, \mathbb{C})$ with real form $G \cong SO(1, n + 1)$, given by the involution $\tau(X) = T_0 \bar{X} T_0$, where T_0 is the diagonal matrix with diagonal entries $-1, 1, 1, \ldots$ and the symmetric space structure is given by the involution $\sigma(B) = SBS$, where S is a diagonal matrix, with the first four entries equal to -1 and all other entries equal to 1.

Thus $K^{\mathbb{C}} = SO(4, C) \times SO(n - 2, \mathbb{C})$. The general theory we have presented in this paper does allow for many open Iwasawa cells and we believe that they actually exist and will give rise to many effects for general Willmore surfaces in S^n. (We have not yet computed all Iwasawa cells in this example.)

However, fortunately, for the case of Willmore spheres things are very simple. It turns out [8] that in this case the corresponding conformal harmonic maps are of 'finite uniton number' and therefore the corresponding normalized potentials can be represented as upper triangular matrices. As a consequence, the Iwasawa splittings can be carried out explicitly and it seems that in this case of Willmore spheres the matrix function $C(z, \lambda)$ always stays in the open Iwasawa cells which contains I. Hence the problems discussed in Example 1 will not occur for Willmore spheres (but possibly for Willmore tori).

Remark 4.6.2 Investigating CMC surfaces in $\mathbb{H}^3$ with mean curvature H satisfying $-1 < H < 1$, see [6], one encounters a situation which is similar to the one encountered in this note, but definitely different: the involution τ used in [6] is not induced from an involution of some finite dimensional Lie group (hence it is an intrinsically loop group involution). In particular, neither the work of Kellersch nor any result of this note can be applied directly. Nevertheless one can compute all Iwasawa cells explicitly and obtains exactly two open Iwasawa cells. The behaviour of immersions near the boundary is analogous to Example 1. (also see [12]).

References

[1] K. Aomoto, On some double coset decompositions of complex semisimple Lie groups, J.Math.Soc.Japan **18** (1966), 1–44

[2] D. Brander, W. Rossman, N. Schmitt, Holomorphic representation of constant mean curvature surfaces in Minkowski space:consequences of non-compactness in loop group methods, Adv.Math. **223** (2010), 949–986

[3] J. Dorfmeister, H. Gradl, J. Szmigielski, Systems of PDEs obtained from factorization in loop groups, Acta Appl.Math. **53** (1998), 1–58

[4] J. Dorfmeister, Weighted l_1 −Grassmannians and Banach manifolds of solutions of the KP-equation and the KdV-equation, Math.Nachr. **180** (1996), 43–73

[5] J. Dorfmeister, M. Guest, W. Rossman, The tt^* structure of the quantum cohomology of $\mathbb{C}P^1$ from the viewpoint of differential geometry, Asian J.Math. **14** (2010), 417–438

[6] J. Dorfmeister, J. Inoguchi, S. Kobayashi, Constant mean curvature surfaces in hyperbolic 3-space via loop groups, to appear

[7] J. Dorfmeister, F. Pedit and H. Wu, Weierstrass-type representations of harmonic maps into symmetric spaces, Comm.Anal.Geom. **6** (1998), 633–668.

[8] J. Dorfmeister, P. Wang Willmore Spheres in S^n, work in progress

[9] V. Kac, D. Peterson, Infinite flag varieties and conjugacy theorems, Proc.Nat.Acad.Sci.USA **80** (1983), 1772–1782

[10] P. Kellersch, Eine Verallgemeinerung der Iwasawa Zerlegung in Loop Gruppen, Dissertation, TU München, 1999

[11] P. Kellersch, The Iwasawa decomposition for the untwisted group of loops in semisimple Lie groups Balkan Press 2003, http://www.mathem.pub.ro/dgds/mono/dgdsmono.htm

[12] S. Kobayashi, Totally symmetric surfaces of constant mean curvature in hyperbolic 3-space, Bull.Aust.Math.Soc **82** (2010), 240–253

[13] T. Matsuki, The orbits of affine symmetric spaces under the action of minimal parabolic subgroups, J.Math.Soc.Japan **31** (1979) 331–357

[14] A. Pressley, G. Segal, Loop groups, Oxford Mathematical Monographs, Oxford University Press 1986

[15] W. Rossmann, The structure of semisimple symmetric spaces, Canad.J.Math **31** (1979), 157–180

[16] G. Segal, G. Wilson, Loop groups and equations of KdV type, Inst.Hautes Etudes Sci.Publ.Math. **61** (1985), 5–65

Authors' addresses:

Josef F. Dorfmeister
Fakultät für Mathematik,
Technische Universität München,
Boltzmannstr. 3,
D-85747 Garching,
Germany
dorfm@ma.tum.de

5

Multiplier ideal sheaves and geometric problems

AKITO FUTAKI AND YUJI SANO

Abstract

In this expository article we first give an overview on multiplier ideal sheaves and geometric problems in Kählerian and Sasakian geometries. Then we review our recent results on the relationship between the support of the subschemes cut out by multiplier ideal sheaves and the invariant whose non-vanishing obstructs the existence of Kähler-Einstein metrics on Fano manifolds.

5.1 Introduction

One of the main problems in Kählerian and Sasakian geometries is the existence problem of Einstein metrics. An obvious necessary condition for the existence of a Kähler-Einstein metric on a compact Kähler manifold M is that the first Chern class $c_1(M)$ is negative, zero or positive since the Ricci form represents the first Chern class. This existence problem in Kählerian geometry was settled by Aubin [1] and Yau [58] in the negative case and by Yau [58] in the zero case. In the remaining case when the manifold has positive first Chern class, in which case the manifold is called a Fano manifold in algebraic geometry, there are two known obstructions. One is due to Matsushima [29] which says that the Lie algebra $\mathfrak{h}(M)$ of all holomorphic vector fields on a compact Kähler-Einstein manifold M is reductive. The other one is due to the first author [16] which is given by a Lie algebra character $F : \mathfrak{h}(M) \to \mathbb{C}$ with the property that if M admits a Kähler-Einstein metric then F vanishes identically. Besides, it has been conjectured by Yau [59] that a more subtle condition related to geometric invariant theory (GIT) should be equivalent to the existence of Kähler-Einstein metrics. This idea was made explicit in the paper [50] of Tian in which a notion called K-stability was introduced. Tian used a

68

generalized version of the invariant F for normal almost Fano varieties and used it as the numerical invariant for the stability condition. The link between the idea of GIT stability and geometric problems such as the existence problems of Hermitian-Einstein metrics and constant scalar curvature Kähler metrics can be explained through the moment maps in symplectic geometry. The explanation from this viewpoint can be found for example in [14], [15], [13]. Recall that an extremal Kähler metric is by definition a Kähler metric such that the gradient vector field of the scalar curvature is a holomorphic vector field. In particular, a Kähler metric of constant scalar curvature is an extremal Kähler metric. The theorem of Matsushima is extended for extremal Kähler manifolds by Calabi [3] as a structure theorem of the Lie algebra $\mathfrak{h}(M)$ on an extremal Kähler manifold M, and the first author's obstruction F can be extended as an obstruction to the existence of constant scalar curvature Kähler metric in a fixed Kähler class ([17], [3]). The theorem of Calabi and the character F are explained in the framework of the moment map picture by X. Wang [55] (see also [19]).

In [10] Donaldson refined the notion of K-stability for a polarized manifold (M, L), that is, a pair of an algebraic manifold M and an ample line bundle L over M, and conjectured that there would exist a Kähler form in $c_1(L)$ of constant scalar curvature if and only if (M, L) is K-polystable. To define K-(poly)stability for (M, L) Donaldson refined the invariant F even for non-normal varieties which are degenerations of the polarized manifold (M, L) and used it as the numerical invariant for the stability condition. The notion of K-stability is defined as follows. For an ample line bundle L over a projective variety M of dimension m, a test configuration of exponent r consists of the following:

(1) A flat family of schemes $\pi : \mathcal{M} \to \mathbb{C}$.
(2) A $\mathbb{C}^*$-action on $\mathcal{M}$ covering the usual $\mathbb{C}^*$-action on $\mathbb{C}$.
(3) A $\mathbb{C}^*$-equivariant line bundle $\mathcal{L} \to \mathcal{M}$ such that

- for $t \neq 0$ one has $M_t = \pi^{-1}(t) \cong M$ and $(M_t, \mathcal{L}|_{M_t}) \cong (M, L^r)$,
- $\chi(M_t, L_t^r) = \sum_{p=0}^{m}(-1)^p \dim H^p(M_t, L_t^r)$ does not depend on t, in particular for r sufficiently large $\dim H^0(M_t, L_t^r) = \dim H^0(M, L^r)$ for all $t \in \mathbb{C}$. Here we write L_t^r for $\mathcal{L}|_{M_t}$ though L may not exist for $t = 0$.

The $\mathbb{C}^*$-action on $(\mathcal{L}, \mathcal{M})$ induces a $\mathbb{C}^*$-action on the central fiber $L_0 \to M_0 = \pi^{-1}(0)$. Moreover if (M, L) admits a $\mathbb{C}^*$-action, then one obtains a test configuration by taking the direct product $M \times \mathbb{C}$. This is called a product configuration. A product configuration is called a trivial configuration if the action of $\mathbb{C}^*$ on M is trivial.

Definition 5.1.1 (M, L) is said to be K-semistable (resp. stable) if the invariant F_1 (defined below) of the central fiber (M_0, L_0) is non-positive (resp. negative) for all non-trivial test configurations. (M, L) is said to be K-polystable if it is K-semistable and $F_1 = 0$ only if the test configuration is product.

Here the invariant F_1 is defined as follows. Let $L_0 \to M_0$ be an ample line bundle over an m-dimensional projective scheme. We assume that a $\mathbb{C}^*$-action as bundle isomorphisms of L_0 covers the $\mathbb{C}^*$-action on M_0. For any positive integer k, there is an induced $\mathbb{C}^*$ action on $W_k = H^0(M_0, L_0^k)$. Put $d_k = \dim W_k$ and let w_k be the weight of $\mathbb{C}^*$-action on $\wedge^{d_k} W_k$. For large k, d_k and w_k are polynomials in k of degree m and $m + 1$ respectively by the Riemann-Roch and the equivariant Riemann-Roch theorems. For sufficiently large k we expand

$$\frac{w_k}{k d_k} = F_0 + F_1 k^{-1} + F_2 k^{-2} + \cdots.$$

When M_0 is smooth, F_1 coincides with $F(X)$ up to a negative multiple constant where X is the infinitesimal generator of the $\mathbb{C}^*$-action on M_0.

The necessity of K-polystability for the existence of constant scalar curvature Kähler metric has been studied by Chen and Tian [6], Paul and Tian [33], [34], Donaldson [11], Stoppa [47] and Mabuchi [28].

Returning to Fano manifolds, there are two hopeful approaches to prove the sufficiency of K-polystability for the existence of Kähler-Einstein metrics. One is the Monge-Ampère equation and the other is the Kähler-Ricci flow. In both cases the difficulty arises in the C^0-estimate, and when the C^0-estimate fails the multiplier ideal sheaves and the subschemes cut out by them appear. The multiplier ideal sheaves arising from the Monge-Ampère equation were studied by Nadel [30], and those arising from Ricci flow were studied for example by Phong, Sesum and Sturm [37] and Rubinstein [40]. We will give an overview of this subject in section 2. On the other hand, for a given subscheme V in M, Ross and Thomas [38] considered the test configuration obtained by blowing up $M \times \mathbb{C}$ along $V \times \{0\}$. M is said to be *slope stable* if the invariant F_1 for the test configuration is negative for any V. These works lead us to ask how the invariant F (or more generally F_1) is related to the multiplier ideal sheaves arising from the Monge-Ampère equation and the Ricci flow. We will treat this subject in section 3.

Now we turn to Sasakian geometry. For general facts about Sasakian geometry, refer to the book [2]. Let (S, g) be a Riemannian manifold. We denote its Riemannian cone $(\mathbb{R}_+ \times S, dr^2 + r^2 g)$ by $(C(S), \bar{g})$. A Riemannian manifold (S, g) is said to be a **Sasaki manifold** if the Riemannian cone $(C(S), \bar{g})$ is Kähler. From this definition the dimension of Sasaki manifold (S, g) is odd, and we put $\dim_{\mathbb{R}} = 2m + 1$ so that $\dim_{\mathbb{C}} C(S) = m + 1$. (S, g) is isometric to the submanifold $\{r = 1\} = \{1\} \times S \subset (C(S), \overline{g})$, and we identify S with the submanifold $\{r = 1\}$. Let J be the complex structure on $C(S)$ giving the Kähler structure. Consider the vector field

$$\widetilde{\xi} = Jr\frac{\partial}{\partial r}.$$

Then $\frac{1}{2}(\widetilde{\xi} - iJ\widetilde{\xi})$ is a holomorphic vector field. The restriction ξ of $\widetilde{\xi}$ to $S \cong \{r = 1\}$ becomes a Killing vector field, called the Reeb vector field. The flow generated by ξ is called the Reeb flow. The restriction η, of the 1-form $\widetilde{\eta}$ on $C(S)$ defined as

$$\widetilde{\eta} = \frac{1}{r^2}\overline{g}(\widetilde{\xi}, \cdot) = \sqrt{-1}(\overline{\partial} - \partial)\ln r,$$

to $S \cong \{r = 1\}$ becomes a contact form. Hence $d\eta$ defines Kähler forms on local orbit spaces of Reeb flow. That is to say, the 1-dimensional foliation defined by ξ comes equipped with a structure of transverse Kähler foliation. A Sasaki manifold is said to be regular if the Reeb flow generates a free S^1-action, quasi-regular if all the orbits are closed. A Sasaki manifold is said to be irregular if it is not quasi-regular.

For a polarized manifold (M, ω) the associated $U(1)$-bundle S of L becomes a regular Sasaki manifold in a natural way: Choose a positive $(1, 1)$ form representing $c_1(L)$, take the Hermitian metric h on L such that the connection form $\widetilde{\eta}$ on L has its curvature form $d\widetilde{\eta}$ equal to ω. The Kähler cone $C(S)$ is L minus the zero section with the Kähler form given by $\frac{i}{2}\partial\overline{\partial}r^2$ where r is the distance from the zero section. Conversely, any regular Sasaki manifold is given in this way. Similarly, a quasi-regular Sasaki manifold is given as an associated $U(1)$-orbibundle over an orbifold.

As is shown in [20] most of ideas in Kähler geometry can be extended to transverse Kähler geometry for Sasaki manifolds. For example one can extend Calabi's theorem to compact Sasaki manifold whose transverse Kähler metric is an extremal Kähler metric, and one can extend the obstruction F as an obstruction for a basic cohomology class to admit a transverse Kähler form with constant scalar curvature.

A Sasaki-Einstein manifold is a Sasaki manifold whose metric is an Einstein metric. This condition is equivalent to the transverse Kähler metric being Kähler-Einstein. Thus the study of the existence problem of Sasaki–Einstein metrics are closely related to the problem of Kähler–Einstein metrics. But there are differences between them. To explain the differences let k be the maximal dimension of the torus which acts on $C(S)$ by holomorphic isometries. When $k = m + 1$ the cone $C(S)$ is a toric variety, and in this case the Sasaki manifold S is said to be toric. Notice that k is at least 1 because $\widetilde{\xi}$ generates holomorphic isometries on $C(S)$. The other extreme case is therefore when $k = 1$. In this case the Sasaki manifold is necessarily quasi-regular.

The contact bundle $D = \operatorname{Ker} \eta \subset TS$ has a complex structure given by the restriction of J. A necessary condition for the existence of Sasaki-Einstein metric in a fixed transverse Kähler structure is that the following two conditions are satisfied:

(a) the basic first Chern class is represented by a positive transverse $(1, 1)$-form,

(b) $c_1(D) = 0$,

see [2] or [20] for the proof. In [20] it is proved that if a compact toric Sasaki manifold satisfies the conditions (a) and (b) then we can deform the Reeb vector field so that the resulting Sasaki manifold has a Sasaki-Einstein metric. It can be shown that the conditions (a) and (b) can be rephrased as the Sasaki manifold is obtained from the toric diagram of constant height, and equivalently as the apex of $C(S)$ is a $\mathbb{Q}$-Gorenstein singularity (c.f. [8]). In the case of the other extreme when $k = 1$ the conditions (a) and (b) only say that the orbit space of the Reeb flow is a Fano orbifold. In this case there is no deformation space of Reeb vector field and the problem has the same difficulty as the problem of Kähler-Einstein metrics. For the intermediate cases when $1 < k < m + 1$ the authors do not even know how to state the conjecture. In the extreme case where $k = 1$ numerous existence results were obtained by Boyer, Galicki, Kollár and their collaborators using the multiplier ideal sheaves, which will be reviewed in the next section.

5.2 An overview of multiplier ideal sheaves

In this section, we recall the results about the relationships between the existence of Kähler-Einstein metrics on Fano manifolds and multiplier ideal sheaves, and related topics. In particular we focus on Nadel's works and recent results about multiplier ideal sheaves and the Kähler-Ricci flow.

Nadel [30] gave a sufficient condition for the existence of Kähler-Einstein metrics on Fano manifolds by using multiplier ideal sheaves, which was originally studied in the works of J.J. Kohn. Let M be a compact m-dimensional Fano manifold. Let g be a Kähler metric on M, whose Kähler class equals to the first Chern class $c_1(M)$ of M. Let $\gamma_0 \in (0, 1)$. We denote the Kähler form and the Ricci form of g by ω_g and $\operatorname{Ric}(g)$ respectively. Let

$$S := \{\varphi_k \in C^\infty_{\mathbb{R}}(M) \mid g_{i\bar{j}} + \partial_i \partial_{\bar{j}} \varphi_k > 0, \ \sup_M \varphi_k = 0, \ 1 \leq k < \infty\} \quad (5.1)$$

be a sequence of Kähler potentials with respect to ω_g such that

$$\lim_{k \to \infty} \int_M e^{-\gamma \varphi_k} dV = \infty \tag{5.2}$$

for any $\gamma \in (\gamma_0, 1)$, and that there is a nonempty open subset $U \subset M$ satisfying

$$\int_U e^{-\varphi_k} dV \leq O(1) \tag{5.3}$$

as $k \to \infty$, where dV is a fixed volume form. Note that the last condition (5.3) always holds for any S due to $g_{i\bar{j}} + \partial_i \partial_{\bar{j}} \varphi > 0$ (see for instance [49]). For each S, Nadel constructed a coherent ideal sheaf $\mathcal{I}(S)$, which is called the multiplier ideal sheaf (MIS). We will explain later the simpler definition of multiplier ideal sheaves given by Demailly–Kollár [9].

Here let us recall the outline of Nadel's construction. (See the original paper [30] for the full details.) Let L be an arbitrary ample line bundle on M which is not necessarily the anticanonical line bundle of M. We define $H^0(M, L^\nu)_S$ to be the set of all $f \in H^0(M, L^\nu)$ for which there exists a sequence $\{f_k\}$ of $H^0(M, L^\nu)$ such that

$$\int_M |f_k|^2 e^{-\gamma \varphi_k} dV \leq C$$

for some $\gamma \in (\gamma_0, 1)$ and $f_k \to f$ uniformly. Consider the homogeneous coordinate ring

$$R(M, L) = \bigoplus_{\nu=0}^{\infty} H^0(M, L^\nu).$$

Define

$$I(M, S, L) = \bigoplus_{\nu=0}^{\infty} H^0(M, L^\nu)_S,$$

which is a homogeneous ideal $I(M, S, L)$ of the graded ring $R(M, L)$. Then, the ideal sheaf $\mathcal{I}(S)$ is defined as the algebraic sheaf of ideals on M associated to $I(M, S, L)$. It is proved in [30] that this construction is independent of the choice of L. Let $\mathcal{V}(S)$ be the (possibly non-reduced) subscheme in M cut out by $\mathcal{I}(S)$. This subscheme is characterized as follows. A point $p \in M$ is contained in the complement of $\mathcal{V}(S)$ if and only if there exist an open neighborhood W of p in M and a real number $\gamma \in (\gamma_0, 1)$ such that

$$\int_W e^{-\gamma \varphi_k} dV \leq O(1)$$

as $k \to \infty$. Note that $\mathcal{V}(S)$ is neither empty nor M if S satisfies the conditions (5.1), (5.2) and (5.3).

One of the distinguished properties of this ideal sheaf is the following vanishing theorem.

Theorem 5.2.1 *[30] For every semi-positive Hermitian holomorphic line bundle L on M,*

$$H^i(M, \mathcal{O}(L) \otimes \mathcal{I}(S)) = 0, \text{ for all } i > 0.$$

In particular, we have

$$H^i(M, \mathcal{I}(S)) = 0, \ i \geq 0. \tag{5.4}$$

Remark that (5.4) at $i = 0$ follows from the fact that the subscheme $\mathcal{V}(S)$ is not empty. We also find that (5.4) implies

$$H^0(\mathcal{V}(S), \mathcal{O}_{\mathcal{V}(S)}) = \mathbb{C}, \\ H^i(\mathcal{V}(S), \mathcal{O}_{\mathcal{V}(S)}) = 0 \quad \text{for all } i > 0. \tag{5.5}$$

This vanishing formula (5.5) gives us several geometric properties of $\mathcal{V}(S)$. For example,

(a) $\mathcal{V}(S)$ is connected.
(b) If $\mathcal{V}(S)$ is zero dimensional, then it is a single reduced point.
(c) If $\mathcal{V}(S)$ is one dimensional, then it is a tree of smooth rational curves.

The main result in [30] is that if a Fano manifold does not admit Kähler-Einstein metrics then the bubble of the solution of the continuity method induces a proper multiplier ideal sheaf with the above vanishing formula (5.5). To explain it, let us recall the continuity method for the Monge-Ampère equation. Here, we assume that $\gamma_0 = m/(m + 1)$. Consider the following equation

$$(\det(g_{i\bar{j}}) + \partial_i\partial_{\bar{j}}\varphi_t)/(\det(g_{i\bar{j}})) = \exp(h_g - t\varphi_t), \tag{5.6}$$

where $t \in [0, 1]$ and h_g is the real-valued function defined by

$$\text{Ric}(g) - \omega_g = \frac{\sqrt{-1}}{2\pi}\partial\bar{\partial}h_g, \ \int_M e^{h_g}\omega_g^m = \int_M \omega_g^m = V. \tag{5.7}$$

It is well-known that the space $T := \{t \in [0, 1] \mid (5.6) \text{ has a solution}\}$ contains 0 (due to the Calabi-Yau theorem) and open in $[0, 1]$ (due to the implicit function theorem). If T is closed then (5.6) is solvable at $t = 1$, i.e., $\omega_g + \frac{\sqrt{-1}}{2\pi}\partial\bar{\partial}\varphi_1$ gives a Kähler-Einstein form. A priori estimates for the closedness of T were given by Yau [58]; if (5.6) is solvable at $s \in [0, t)$ and $\|\varphi_s\|_{C^0}$ is uniformly bounded then (5.6) is solvable at $s = t$. Nadel proved that if the solution $\{\varphi_t\}_{0 \leq t < t_0}$ of

(5.6) violates the above estimate, then there is a sequence $\{t_k\}$ such that $t_k \to t_0$ as $k \to \infty$ and $\{\varphi_{t_k} - \sup_M \varphi_{t_k}\}_{k=1}^{\infty}$ induces a proper multiplier ideal sheaf $\mathcal{I}$. In this paper, we call it the **Kähler-Einstein multiplier ideal sheaf** (KE-MIS). Summing up,

Theorem 5.2.2 ([30]) *Let M be a Fano manifold which does not admit a Kähler-Einstein metric. Let G be a compact subgroup of the group $\mathrm{Aut}(M)$ of holomorphic automorphisms of M. Assume that M does not admit any $G^{\mathbb{C}}$-invariant proper multiplier ideal sheaf. Then M admits Kähler-Einstein metrics. Here $G^{\mathbb{C}}$ denotes the complexification of G.*

By combining the above theorem and the geometric properties of $\mathcal{V}(S)$ given by the vanishing formula (5.5), Nadel gave many examples of Kähler-Einstein Fano manifolds. Recently Heier [23] applied this method to (re-)prove the existence of Kähler-Einstein metrics on complex del Pezzo surfaces obtained from the blow up of $\mathbb{CP}^2$ at 3,4 or 5 points, which was originally proved by Siu [44], Tian [49], Tian and Yau [51].

This method was extended to the case of Fano orbifolds by Demailly-Kollár [9]. Their construction is simpler than [30]. Let ψ be an ω_g-plurisubharmonic (psh) function (or almost psh function with respect to ω_g), i.e., a real-valued upper semi-continuous function satisfying $\omega_g + \frac{\sqrt{-1}}{2\pi}\partial\bar{\partial}\psi \geq 0$ in the current sense. The multiplier ideal sheaf with respect to ψ in the sense of [9] is the ideal sheaf defined by the following presheaf

$$\Gamma(U, \mathcal{I}(\psi)) = \{f \in \mathcal{O}(U) \mid \int_U |f|^2 e^{-\psi} dV < \infty\} \tag{5.8}$$

where U is an open subset of M. This sheaf is also coherent and satisfies the vanishing theorem of Nadel type. In terms of this formulation, Theorem 5.2.2 can be written as follows. Let $\{\varphi_t\}$ be the solution $\{\varphi_t\}_{0 \leq t < t_0}$ of (5.6) which violates a priori estimates.

Theorem 5.2.3 ([9]) *Let M be a Fano manifold of dimension m. Let G be a compact subgroup of $\mathrm{Aut}(M)$. Assume that M does not admit a G-invariant Kähler-Einstein metric. Let $\gamma \in (m/(m+1), 1)$. Then there exists a G-invariant sequence $\{\varphi_{t_k}\}_{k=1}^{\infty}$ such that*

- *$t_k \to t_0$ as $k \to \infty$,*
- *there exists a limit $\varphi_{\infty} = \lim_{k \to \infty}(\varphi_{t_k} - \sup_M \varphi_{t_k})$ in the L^1-topology, which is an ω_g-psh function, and*
- *$\mathcal{I}(\gamma\varphi_{\infty})$ is a $G^{\mathbb{C}}$-invariant proper multiplier ideal sheaf, i.e, $\mathcal{I}(\gamma\varphi_{\infty})$ is neither 0 nor $\mathcal{O}_M$.*

We call $\mathcal{I}(\gamma\varphi_\infty)$ the KE-MIS of exponent γ. In the above, one of the important ingredients is that the upper bound of γ is strictly smaller than 1. To explain this point, we shall state Nadel's vanishing theorem in terms of Demailly-Kollár's formulation. Let L be a holomorphic line bundle over M with a singular Hermitian metric $h = h_0 e^{-\psi}$, where h_0 is a smooth Hermitian metric and ψ is a L^1_{loc}-function. Assume that $\Theta_h(L) = \frac{\sqrt{-1}}{2\pi}\partial\bar{\partial}(-\ln h_0 + \psi)$ is positive definite in the sense of currents, i.e., $\Theta_h(L) \geq \varepsilon\omega_g$ for some $\varepsilon > 0$. Then, in the same spirit of Nadel's vanishing theorem, we have

$$H^i(M, K_M \otimes L \otimes \mathcal{I}(\psi)) = 0, \quad i > 0, \tag{5.9}$$

where K_M is the canonical line bundle. Now let φ be an ω_g-psh function on a Fano manifold M with $[\omega_g] = c_1(M)$. Substitute K_M^{-1} and $\gamma\varphi$ into L and ψ in (5.9) respectively, and assume h_0 associates to ω_g. Since $\omega_\varphi = \omega_g + \frac{\sqrt{-1}}{2\pi}\partial\bar{\partial}\varphi \geq 0$, we have

$$\Theta_{h_0 e^{-\gamma\varphi}}(L) = \gamma\omega_\varphi + (1-\gamma)\omega_g \geq (1-\gamma)\omega_g$$

if $\gamma < 1$. This means that the positivity condition for (5.9) with respect to $h_0 e^{-\gamma\varphi}$ holds if $\gamma < 1$. Then (5.9) implies

$$H^i(M, \mathcal{I}(\gamma\varphi)) = 0, \quad i > 0.$$

Moreover, if the subscheme cut out by $\mathcal{I}(\gamma\varphi)$ is not empty, then

$$H^0(M, \mathcal{I}(\gamma\varphi)) = 0.$$

Summing up, we get

Lemma 5.2.4 *If there exists a positive constant $\gamma < 1$ and an ω_g-psh function φ such that $\mathcal{I}(\gamma\varphi)$ is proper, then $\mathcal{I}(\gamma\varphi)$ satisfies (5.4). In particular, $\mathcal{I}(\gamma\varphi_\infty)$ for $\gamma \in (m/(m+1), 1)$ in Theorem 5.2.3 satisfies (5.4) (and then (5.5)).*

On the other hand, the lower bound of γ in Theorem 5.2.3 describes the strength of the singularity of φ_∞. It is closely related to a holomorphic invariant introduced by Tian [49]. It is often called the α-invariant, which is defined by

$$\alpha_G(M) := \sup\{\alpha \in \mathbb{R} \mid \int_M e^{-\alpha(\psi - \sup_M \psi)}\omega_g^m < C_\alpha \text{ for all } G\text{-invariant } \omega_g\text{-psh } \psi\} \tag{5.10}$$

where $G \subset \mathrm{Aut}(M)$ is a compact subgroup. If a multiplier ideal sheaf $\mathcal{I}(\gamma\psi)$ with respect to a G-invariant ω_g-psh function ψ of exponent γ is proper, where $\sup_M \psi = 0$, then $\alpha_G(M) \leq \gamma$, because $e^{-\gamma\psi}$ is not integrable over M. Conversely,

Lemma 5.2.5 *If $\alpha_G(M) < 1$, then there exist a positive constant $\gamma \in (0, 1)$ and a G-invariant ω_g-psh function ψ with $\sup_M \psi = 0$ such that $\mathcal{I}(\gamma \psi)$ is proper.*

Tian gave a sufficient condition for the existence of Kähler-Einstein metrics on Fano manifolds in terms of this invariant.

Theorem 5.2.6 ([49]) *If $\alpha_G(M) > m/(m + 1)$, then M admits a G-invariant Kähler-Einstein metric.*

Using Theorem 5.2.6, Tian and Yau [51] proved the existence problem of Kähler-Einstein metrics on Fano surfaces, i.e., the Fano surfaces obtained from the blow up of $\mathbb{CP}^2$ at k points where $3 \leq k \leq 8$ admits a Kähler-Einstein metric. Both of the lower bound of $\alpha_G(M)$ and the non-existence of the proper multiplier ideal sheaves satisfying (5.5) give sufficient condition for the existence of Kähler-Einstein metrics on Fano manifolds, and they are related directly to each other. For example, Lemma 5.2.4 and 5.2.5, we have

Lemma 5.2.7 *If $\alpha_G(M) < 1$, then a $G^{\mathbb{C}}$-invariant proper multiplier ideal sheaves satisfying (5.5) exists.*

Although $\alpha_G(M)$ is difficult to compute in general, it is possible to calculate it when M has a large symmetry in such cases as [46] for toric varieties and [12] for the Mukai-Umemura 3-folds. On the other hand, there is a local version of the $\alpha_G(M)$-invariant, which is called the **complex singularity exponent** [9]. Let $K \subset M$ be a compact subset and ψ be a G-invariant ω_g-psh function on M. Then the complex singularity exponent $c_K(\psi)$ of ψ with respect to K is defined by

$$c_K(\psi) = \sup\{c \geq 0 \mid e^{-c\psi} \text{ is } L^1 \text{ on a neighborhood of } K\}.$$

This constant depends only on the singularity of ψ near K. It is obvious that $c_K(\psi) \geq \alpha_G(M)$. One of the important properties of $c_K(\psi)$ is the semi-continuity with respect ψ. Let $\mathcal{P}(M)$ be the set of all locally L^1 ω_g-psh functions on M with L^1-topology. Then, we have (cf. Effective version of Main Theorem 0.2 in [9])

Theorem 5.2.8 ([9]) *Let $K \subset M$ be a compact subset of M. Let $\varphi \in \mathcal{P}(M)$ be given. If $c < c_K(\varphi)$ and $\psi_j \to \varphi$ in $\mathcal{P}(M)$ as $j \to \infty$, then $e^{-c\psi_j} \to e^{-c\varphi}$ in L^1-norm over some neighborhood U of K.*

In particular, if $\{\psi_j\}$ satisfies

$$\int_M e^{-\gamma \psi_j} dV \to \infty$$

where $\gamma \in (\gamma_0, 1)$ and $\psi_j \to \varphi$ in $\mathcal{P}(M)$, then $c_M(\varphi) \leq \gamma_0$. This theorem allows us to substitute Theorem 5.2.3 for Theorem 5.2.2. In fact, if the solution φ_t of (5.6) violates a priori C^0-estimates at $t = t_0$, by using a Harnack inequality we can show

$$\int_M e^{-\gamma(\varphi_t - \sup \varphi_t)} dV \to \infty \text{ as } t \to t_0$$

for any $\gamma \in (m/(m+1), 1)$. In Theorem 5.2.2, a subsequence of $\{\varphi_t\}$ induces the KE-MIS, which is proper. On the other hand, Theorem 5.2.8 implies that $e^{-\gamma \varphi_\infty}$ is not integrable over M for any $\gamma \in (m/(m+1), 1)$, where $\varphi_\infty :=$ $\lim_{i \to \infty}(\varphi_{t_i} - \sup \varphi_{t_i})$. This means that φ_∞ induces the KE-MIS in Theorem 5.2.3.

The multiplier ideal sheaves in [9] and the complex singularity exponent can be defined algebraically as follows (cf. [27] and [2] for instance). Here we consider a smooth variety M of dimension m. Let $D = \sum a_i D_i$ be a $\mathbb{Q}$-divisor on M. A *log resolution* of (M, D) is a projective birational map $\mu : M' \to M$ with M' smooth such that the divisor

$$\mu^* D + \sum_i E_i$$

has simple normal crossing support. Assume D is effective and fix a log resolution μ of (M, D). Then the multiplier ideal sheaf $\mathcal{I}(M, D) \subset \mathcal{O}_M$ with respect to D is defined by

$$\mu_* \mathcal{O}_{M'}(K_{M'} - \mu^*(\lfloor K_M + D \rfloor)), \tag{5.11}$$

where $\lfloor K_M + D \rfloor$ means the integral part of $K_M + D$. Note that $\mathcal{I}(M, D)$ is independent of the choice of μ. This (algebraic) ideal sheaf corresponds to the following (analytic) multiplier ideal sheaf defined in [9]. Take an open set $U \subset M$ so that for each D_i there is a holomorphic function g_i locally defining D_i in U. Let $\varphi_D := \sum_i 2a_i \ln |g_i|$ which is plurisubharmonic on U and define

$$\Gamma(U, \mathcal{I}(\varphi_D)) := \{ f \in \mathcal{O}_M(U) \mid \frac{|f|^2}{\prod |g_i|^{2a_i}} \in L^1_{\text{loc}} \} \tag{5.12}$$

as before. For simplicity, we assume that $D = \sum_i a_i D_i$ has simple normal crossing support. The holomorphic function f satisfies the L^2-integrability condition in (5.12) if and only if f can be divided by $\prod g^{m_i}$ where $m_i \geq \lfloor a_i \rfloor$, i.e., $\mathcal{I}(\varphi_D) = \mathcal{O}_M(-\lfloor D \rfloor)$. Let $\mu : M' \to M$ be a log resolution of D. Then we have

$$\mathcal{I}(M, D) = \mu_* \mathcal{O}_{M'}(K_{M'} - \mu^*(\lfloor K_M + D \rfloor)) = \mathcal{O}_M(-\lfloor D \rfloor) = \mathcal{I}(\varphi_D).$$

The second equality in the above was proved in Lemma 9.2.19 [27]. We also have

$$\mathcal{I}(\varphi_D) = \mathcal{O}_M \iff e^{-\varphi_D} \in L^1 \iff (M, D) \text{ is KLT.} \tag{5.13}$$

Here we say that a pair (M, D) is KLT if and only if

$$\text{ord}_E(K_{M'} - \mu^*(\lfloor K_M + D \rfloor)) > -1$$

for every exceptional divisor E with respect to a log resolution $\mu : M' \to M$. In particular (M, D) is KLT is equivalent to that $\mathcal{I}(M, D) = \mathcal{O}_X$. The equivalent relation (5.13) essentially follows from that if D_i is defined by $\{z_i = 0\}$ for a local coordinate $\{z_i\}$ then the L^1-integrability of $e^{-\varphi_D}$ is equivalent to that $a_i < 1$ for all i (Proposition 3.20 [24]). In particular,

$$(M, \gamma D) \text{ is KLT} \iff e^{-\gamma \varphi_D} \in L^1. \tag{5.14}$$

Note that this holds for an (M, D) where D does not necessarily have simple normal crossing support (Proposition 3.20 [24]). By using the KLT condition (5.14), we can rephrase Theorem 5.2.3. Assume that a Fano manifold M does not admit a Kähler-Einstein metric. Let φ_t be the solution of (5.6) where $t \in [0, t_0)$. As explained before, by taking a subsequence of $\{\varphi_{t_j} - \sup_M \varphi_{t_j}\}$, there exists a limit φ_∞ in L^1-topology, which is an ω_g-psh function, such that $e^{-\gamma \varphi_\infty} \notin L^1$ for all $\gamma \in (m/(m+1), 1)$. Since an approximation theorem in [9] implies that any ω_g-psh function can be approximated by an ω_g-psh function formed of $\ln(\sum_i |f_i|^2)$ where all f_i are holomorphic functions, we can replace the above φ_∞ by an ω_g-psh function formed of $\frac{2}{s} \ln |\tau_s|$ where $\tau_s \in H^0(M, K_M^{-s})$ for sufficiently large s. That is to say, there exist a sufficiently large integer s and $\tau_s \in H^0(M, K_M^{-s})$ such that $e^{-2\gamma \frac{1}{s} \ln |\tau_s|} = |\tau_s|^{-\frac{2\gamma}{s}} \notin L^1$ for all $\gamma \in (m/(m+1), 1)$. Here $|\cdot|$ is the induced Hermitian metric on K_M^{-s} with respect to the Kähler metric g. Hence we have the following theorem. Note that the original result holds for orbifolds, but for simplicity we assume that M is smooth in this paper.

Theorem 5.2.9 (Theorem 20 [24], Theorem 5.2.16 [2]) *Let M be a Fano manifold of dimension m and G be a finite subgroup of $\text{Aut}(M)$. Assume that there is an $\varepsilon > 0$ such that a pair $(M, \frac{m+\varepsilon}{m+1} D)$ is KLT for every G-invariant effective divisor D which is numerically equivalent to K_M^{-1}. Then M has a G-invariant Kähler-Einstein metric.*

The complex singularity exponent can be also defined algebraically as follows, which is called the *log canonical thresholds* (cf. Appendix in [5]). Let $Z \subset M$ be a closed subvariety. For an effective $\mathbb{Q}$-Cartier divisor D on M, the log canonical threshold of D along Z is defined by

$$\mathrm{lct}_Z(M, D) := \sup\{\lambda \in \mathbb{Q} \mid \text{the pair } (M, \lambda D) \text{ is log canonical along } Z\}.$$

Here, the pair (M, D) is called log canonical along Z if $\mathcal{I}(M, (1 - \varepsilon)D)$ is trivial in a neighborhood of every point $x \in Z$ for all $0 < \varepsilon < 1$. For instance, let us consider a simple case. Let M be a Fano manifold and $\sigma \in H^0(M, K_M^{-l})$. Let ψ_σ be an ω-psh function defined by $\psi_\sigma(z) = \frac{1}{l} \ln |\sigma(z)|$. Let D_σ be the associated divisor with σ and Z be a closed subvariety in M. In this case, $\mathrm{lct}_Z(M, \frac{1}{l}D)$ is the same as $c_Z(\psi_\sigma)$. The log canonical threshold plays an important role in the studies of the multiplier ideal (sheaves) in algebraic geometry (cf. [27]). Hence we could expect that the complex singularity exponent with respect to the limit φ_∞ in Theorem 5.2.3 has something to do with the existence of Kähler-Einstein metrics although it is not clear at the moment.

To find Kähler-Einstein metrics on Fano manifolds, there is another way instead of solving (5.6), which is the (normalized) Kähler-Ricci flow. The Ricci flow was introduced by R. Hamilton, and on a Fano manifold M with Kähler class $c_1(M)$ it is defined by

$$\frac{d}{dt}\omega_t = -\mathrm{Ric}(\omega_t) + \omega_t, \qquad \omega_0 = \omega_g \tag{5.15}$$

where $t \in [0, \infty)$ and ω_t is the Kähler form of the evolved Kähler metric g_t. Note that (5.15) is normalized so that the Kähler class of g_t is preserved. The existence and uniqueness of the solution of (5.15) for $t \in [0, \infty)$ was proved by Cao [4]. If (5.15) converges in C^∞, the limit is a Kähler–Einstein metric. Then, it is natural to ask whether or not the results about the multiplier ideal sheaves obtained from the continuity method also hold for the Kähler–Ricci flow. The first result of this issue was given by Phong–Sesum–Sturm [37] (see also [36]). The equation (5.15) can be reduced to the equation at the potential level

$$\frac{d}{dt}\varphi_t = \ln(\omega_t^m/\omega_g^m) + \varphi_t - h_g, \quad \varphi_0 = c \tag{5.16}$$

where c is a constant and $\omega_t = \omega_g + \frac{\sqrt{-1}}{2\pi}\partial\bar{\partial}\varphi_t$. They gave a necessary and sufficient condition condition for the convergence of φ_t as $t \to \infty$. Their proof consists of the parabolic analogue of Yau's arguments for the elliptic Monge-Ampère equation, the estimates about the Kähler-Ricci flow by Perelman (cf. [43]) and the result about the Monge-Ampère equations by Kolodziej ([25], [26]).

Theorem 5.2.10 ([37]) *For a certain appropriate constant $c = c_0$, the convergence of the solution of (5.16) is equivalent to that there exists $p > 1$*

such that

$$\sup_{t \geq 0} \frac{1}{V} \int_M e^{-p\varphi_t} \omega_g^m < \infty.$$

The convergence is then in C^∞ and exponentially fast.

To restate the above theorem in terms of the multiplier ideal sheaves, they introduced the sheaf $\mathcal{J}^p$ with respect to a family $\{\psi_t\}_{0 \leq t < \infty}$ of Kähler potentials, which is defined by the presheaf

$$\Gamma(U, \mathcal{J}^p) = \{f \in \mathcal{O}(U) \mid \sup_{t \geq 0} \int_M |f|^2 e^{-p\psi_t} \omega_g^m < \infty\}. \tag{5.17}$$

Hence Theorem 5.2.10 implies the following.

Corollary 5.2.11 ([37]) *The Kähler-Ricci flow converges if and only if there exists $p > 1$ such that $\mathcal{J}^p$ contains the global section 1.*

The sheaf $\mathcal{J}^p$ gives a necessary and sufficient condition for the existence of Kähler-Einstein metrics and the lower bounds of p is optimal (cf. remarks in [37]), whereas such results are not known for the case of the continuity method. We emphasize that the sheaf $\mathcal{J}^p$ contains different information from the multiplier ideal sheaves in Theorem 5.2.3, because we do not need the limit of $\{\varphi_t\}$ but the whole of $\{\varphi_t\}$ in order to define $\mathcal{J}^p$. In fact, in order to get the limit of $\{\varphi_t\}$, we need the appropriate normalization of $\{\varphi_t\}$ as Theorem 5.2.3. In the terminology of the recent paper [45], the $\mathcal{J}^p$ can be regarded as a dynamic MIS which is similar to the Nadel's formulation rather than a static MIS as the Demailly-Kollár's formulation. Now let us consider Theorem 5.2.10 by using the static MIS instead of the dynamic MIS. Theorem 5.2.10 implies that if the normalized Kähler-Ricci flow does not converge, then there is a subsequence $\{\varphi_{t_i}\}_i$ of the solution of (5.16) such that

$$\int_M e^{-p(\varphi_{t_i} - \sup \varphi_{t_i})} \omega_g^m \to \infty$$

as $i \to \infty$ for any $p > 1$. Hence, the limit $\varphi_\infty := \lim(\varphi_{t_i} - \sup \varphi_{t_i})$ implies the multiplier ideal sheaf $\mathcal{I}(p\varphi_\infty)$, which is proper for any $p > 1$. That is to say, if there is no G-invariant ω_g-psh function ψ such that $\mathcal{I}(p\psi)$ is proper for any $p \in (1, +\infty)$, then M admits a G-invariant Kähler-Einstein metric. More precisely,

Theorem 5.2.12 *[37] Let M be a Fano manifold. Let $G \subset \text{Aut}(M)$ be a compact subgroup. Assume that M does not admit Kähler-Einstein metrics. Let $p \in (1, \infty)$ and $\omega_g \in c_1(M)$. There is a G-invariant subsequence of the solutions $\{\varphi_{k_j}\}_{j \geq 1}$ of (5.16) such that*

- *there exists the limit* $\varphi_\infty = \lim_{j\to\infty}(\varphi_{k_j} - \frac{1}{V}\int_M \varphi_{k_j}\omega_g^m)$ *in* L^1-*topology, which is an* ω_g-*psh function, and*
- $\mathcal{I}(p\varphi_\infty)$ *is a* $G^{\mathbb{C}}$-*invariant proper multiplier ideal sheaf satisfying*

$$H^i(M, \mathcal{I}(p\varphi_\infty) \otimes K_M^{-\lfloor p \rfloor}) = 0, \quad \text{for all } i \geq 1.$$

Note that Nadel's vanishing formula (5.5) need not hold for the induced MIS $\mathcal{I}(p\varphi_\infty)$, because $p > 1$. However, this result still has an application. By using a weaker version of Nadel's vanishing theorem and Corollary 5.2.11, Heier [23] proved the existence of Kähler-Einstein metrics for certain del Pezzo surfaces with large automorphism group.

After [37], Rubinstein [40] gave an analogous result as Theorem 5.2.3 for the Kähler-Ricci flow by using a static MIS as Demailly-Kollár. His proof is similar to the case of the continuity method, and makes use of the estimates of Perelman and the uniform Sobolev inequality of the Kähler-Ricci flow given by Ye [60] and Zhang [61], which appeared after [37], in stead of Kolodziej's theorem.

Theorem 5.2.13 ([40])   *Let M be a Fano manifold of dimension m. Let G be a compact subgroup of $\mathrm{Aut}(M)$. Let $\gamma \in (m/(m+1), 1)$. Assume that M does not admit a G-invariant Kähler-Einstein metric. Then there is an initial constant $c_0 = \varphi_0$ and a G-invariant subsequence $\{\varphi_{t_i}\}$ of the solution of (5.16) such that*

- *there exists the limit* $\varphi_\infty = \lim_{t_i\to\infty}(\varphi_{t_i} - \frac{1}{V}\int_M \varphi_{t_i}\omega_g^m)$ *in* L^1-*topology, which is an* ω_g-*psh function, and*
- $\mathcal{I}(\gamma\varphi_\infty)$ *is a* $G^{\mathbb{C}}$-*invariant proper multiplier ideal sheaf.*

In this paper, we call the above multiplier ideal sheaf $\mathcal{I}(\gamma\varphi_\infty)$ the **KRF multiplier ideal sheaf** of exponent γ. There are some remarks about the above theorem. First, the KRF multiplier ideal sheaf is independent of the choice of initial constant c_0 of (5.16) due to the normalization of φ_{t_i}. In fact, if we choose another constant c_0' instead of the constant c_0 in Theorem 5.2.13, which is the same as Theorem 5.2.10, the solution of (5.16) is given by $\varphi_t' = \varphi_t + (c_0' - c_0)e^t$. In contrast to this, when we consider the convergence of non-normalized Kähler potentials $\{\varphi_t\}$ as Theorem 5.2.10, we need to pay attention to the choice of the constant c_0. Second, the normalization in Theorem 5.2.13 is equivalent to the one in Theorem 5.2.3. In fact, there is a uniform constant C such that $\sup_M \varphi_t - C \leq \frac{1}{V}\int_M \varphi_t\omega^m \leq \sup_M \varphi_t$. Third, γ is contained in the interval $(m/(m+1), 1)$. This means that the subscheme cut out by $\mathcal{I}(\gamma\varphi_\infty)$ satisfies (5.5) and we can make use of the induced geometric properties. Fourth, the process to prove Theorem 5.2.13 is similar to the case of the continuity method, and the proof in [40] implies immediately that if $\alpha_G(M) > \frac{m}{m+1}$ then the Kähler-Ricci flow will converge. (This similarity is pointed out in [7] after [40] too.)

Rubinstein [41] also gave the analogous result of Theorem 5.2.12 in terms of the discretization of the Kähler-Ricci flow called "Ricci iteration." Given a Kähler form $\omega \in c_1(M)$ and a real number $\tau > 0$, the time τ Ricci iteration is defined by the sequence $\{\omega_{k\tau}\}_{k \geq 0}$ satisfying

$$\omega_{k\tau} = \omega_{(k-1)\tau} + \tau\omega_{k\tau} - \tau\mathrm{Ric}(\omega_{k\tau}) \text{ for } k \in \mathbb{N}, \quad (5.18)$$

and $\omega_0 = \omega$. When $\tau = 1$, (5.18) is the discretization of (5.15). Let $\mathcal{H}_\omega$ be the space of Kähler potentials with respect to $(\omega, c_1(M))$

$$\mathcal{H}_\omega := \{\psi \in C_\mathbb{R}^\infty \mid \omega_\psi = \omega + \frac{\sqrt{-1}}{2\pi}\partial\bar{\partial}\psi > 0\}.$$

Let h_{ω_ψ} be the Ricci potential with respect to ω_ψ defined as (5.7). Since $[\omega_{k\tau}] = [\omega_{(k-1)\tau}]$, so (5.18) can be written as the system of complex Monge-Ampère equations

$$\omega_{\psi_{k\tau}}^m = \omega^m e^{h_\omega + \frac{1}{\tau}\varphi_{k\tau} - \psi_{k\tau}} = \omega_{\psi_{(k-1)\tau}}^m e^{(\frac{1}{\tau}-1)\varphi_{k\tau} - \frac{1}{\tau}\varphi_{(k-1)\tau}}, \quad (5.19)$$

where $k \in \mathbb{N}$, $\omega_{\psi_{k\tau}} = \omega_{k\tau}$ and $\varphi_{k\tau} := \psi_{k\tau} - \psi_{(k-1)\tau}$.

Theorem 5.2.14 *[41] Let M be a Fano manifold. Let $G \subset Aut(M)$ be a compact subgroup. Assume that M does not admit Kähler-Einstein metrics. Let $\tau = 1$. Let $\gamma \in (1, \infty)$ and $\omega \in c_1(M)$. There is a G-invariant subsequence of the solutions $\{\psi_{k_j}\}_{j \geq 1}$ of (5.19) such that*

- *there exists the limit $\varphi_\infty = \lim_{j \to \infty}(\psi_{k_j} - \frac{1}{V}\int_M \psi_{k_j}\omega^m)$ in L^1-topology, which is an ω-psh function, and*
- *$\mathcal{I}(\gamma\varphi_\infty)$ is a $G^\mathbb{C}$-invariant proper multiplier ideal sheaf satisfying*

$$H^i(M, \mathcal{I}(\gamma\varphi_\infty) \otimes K_M^{-\lfloor\gamma\rfloor}) = 0, \text{ for all } i \geq 1.$$

Considering Yau's conjecture, it is also natural to ask how stability conditions in the sense of GIT is related to the convergence of the Kähler-Ricci flow. For example, see [35], [48] and [54] for references on this issue.

5.3 Direct relationships between multiplier ideal sheaves and the obstruction F

It is conjectured by Yau that the existence of canonical Kähler metrics such as Kähler-Einstein metrics and constant scalar curvature metrics for a given

Kähler class would be equivalent to the stability of manifolds in some sense of Geometric Invariant Theory. This conjecture is formulated by Tian and Donaldson in terms K-polystability as explained in section 1 and is still open.

This conjecture is an analogue of the so-called Hitchin-Kobayashi correspondence, which was proved by Donaldson, and Uhlenbeck and Yau. The proof of the direction from stability towards the existence of Hermitian-Einstein metrics was preceded by constructing subsheaves which violate stability (such sheaves are often called *destabilizing subsheaves*) from the bubble of the Yang-Mills heat flow or the continuity method if a vector bundle does not admit a Hermitian-Einstein metric. Weinkove [57] defined a MIS for each sequence of Hermitian metrics on a holomorphic vector bundle and by using it he proved that the bubble of the Yang-Mills heat flow induces a destabilizing subsheaf. Hence, as the analogy between the Yau-Tian-Donaldson conjecture and the Hitchin-Kobayashi correspondence, we could expect that the MIS obtained from the continuity method or the Kähler-Ricci flow corresponds to a destabilizing subsheaf in some sense for a Fano manifold with anticanonical polarization, but their relation is not clear at this moment. This issue leads us to study direct relationships between multiplier ideal sheaves and the obstruction F. Such a direct relationship was first pointed out by Nadel in [31]. Extending Nadel's result is the main purpose of this section.

Up to this point we have not defined the character F explicitly, which we do now. Let M be an m-dimensional Fano manifold with Kähler class $c_1(M)$ and g be a Kähler metric whose Kähler form ω_g represents $c_1(M)$. We denote the Lie algebra consisting of all holomorphic vector fields on M by $\mathfrak{h}(M)$. We define the map $F : \mathfrak{h}(M) \to \mathbb{C}$ by

$$F(v) := \int_M v h_g \omega_g^m.$$

In [16], the first author proved that F is independent of the choice of g and that F is a Lie algebra character of $\mathfrak{h}(M)$. If M is a Kähler-Einstein manifold, then F vanishes on $\mathfrak{h}(M)$ because we can take $h_g \equiv 0$. Thus the vanishing of F is a necessary condition for the existence of Kähler-Einstein metrics, but it is known that it is not sufficient. For example, in [50] Tian gave a counterexample which does not admit Kähler-Einstein metrics and have no nontrivial holomorphic vector fields. So it is reasonable to study relationships between the invariant F and multiplier ideal sheaves. First of all, we consider the multiplier ideal sheaves obtained from the continuity method in the sense of Nadel. Assume that M does not admit Kähler-Einstein metrics and that $\mathfrak{h}(M) \neq \{0\}$. For each nontrivial holomorphic vector field v, define

$$Z^+(v) := \{p \in \mathrm{Zero}(v) \mid \mathrm{Re}(\mathrm{div}(v)) > 0\},$$

where $\text{Zero}(v)$ is the zero set of v and $\text{div}(v)$ is the divergence of v with respect to some Kähler metric g, i.e., $\text{div}(v) = (\mathcal{L}_v(\omega_g^m))/\omega_g^m$, $\mathcal{L}_v$ being the Lie derivative along v. Note that $Z^+(v)$ does not depend on the choice of g, although $\text{div}(v)$ does. Since M does not admit a Kähler-Einstein metric, the closedness of the set of $t's$ for which the solutions $\{\varphi_t\}$ of (5.6) exist does not hold, that is, the solutions cease to exist at some $t_0 \in (0, 1]$. Then the main result in [31] is as follows.

Theorem 5.3.1 ([31]) *Let M, $\mathfrak{h}(M)$ and $\{\varphi_t\}$ be as above. Let $\mathcal{V}$ be the induced KE-MIS obtained from a subsequence of $\{\varphi_{t_i}\}_i$ where $t_i < t_0$ and $t_i \to t_0$. Then, for any $v \in \mathfrak{h}(M)$ with $F(v) = 0$, the support of $\mathcal{V}$ is not contained in $Z^+(v)$.*

By using the above theorem, Nadel gave another theoretical approach to show that $\mathbb{CP}^1$ does admit a Kähler-Einstein metric. In fact, if we assume that $\mathbb{CP}^1$ did not admit a Kähler-Einstein metric, then $\mathcal{V}$ would be zero dimensional and it would be a single reduced point, which follows from (b) of the properties of the multiplier ideal subschemes. We may assume that $\mathcal{V} = \{z = 0\}$ in $\mathbb{CP}^1 = \mathbb{C} \cup \{\infty\}$. Let $v = z\frac{d}{dz} \in \mathfrak{h}(M)$, then $v = 0$ and the divergence of v is strictly positive at $z = 0$. Hence $\mathcal{V} \subset Z^+(v)$, which is a contradiction. As far as the authors know, other applications of Theorem 5.3.1 except this example had been unknown until [21].

We wish to extend this in several ways. We wish first of all to get some more informations about Fano manifolds, secondly to show the existence of MIS for Kähler-Ricci solitons, and thirdly to study the MIS arising from the non-convergence of Kähler-Ricci flow and study the relation between MIS and F.

We study three types of MIS: first of all KE-MIS which is due to Nadel, arising from the failure of solving Monge-Ampère equations for Kähler-Einstein metrics by continuity method, secondly KRS-MIS which arises from the failure of solving Monge-Ampère equations for Kähler-Ricci solitons by the continuity method and thirdly KRF-MIS which arises from the failure of convergence of Kähler-Ricci flow.

Let M be a Fano manifold, G a compact subgroup of $\text{Aut}(M)$, and T^r the maximal torus of G. For any G-invariant Kähler metric g with

$$\omega_g := \frac{\sqrt{-1}}{2\pi} g_{i\bar{j}} dz^i \wedge d\bar{z}^j \in c_1(M)$$

consider the Hamiltonian T^r-action with the moment map $\mu_g : M \to \mathfrak{t}^{r*}$. We normalize it by

$$\int_M u_X e^h \omega^m = 0$$

where $u_X(p) = \langle \mu(p), X \rangle$ and $\mathrm{Ric}_\omega - \omega = i\partial\bar\partial h$. Note that this normalization is equivalent to requiring u_X to satisfy

$$\Delta u_X + Xh + u_X = 0,$$

see [18]. For $\xi \in \mathfrak{t}^r$ put

$$D^{\leq 0}(\xi) := \{y \in \mu(M) \mid \ < y, \xi > \ \leq 0\}.$$

Theorem 5.3.2 ([21]) *Suppose M does not admit a Kähler-Einstein metric, and let V be the support of the KE-MIS. Let $\xi \in \mathfrak{t}^r \subset \mathfrak{h}(M)$ satisfy $F(v_\xi) > 0$ where v_ξ is the holomorphic vector field corresponding to ξ. Then*

$$\mu_g(V) \not\subset D^{\leq 0}(\xi)$$

for any G-invariant Kähler metric g whose Kähler form is in $c_1(M)$.

Corollary 5.3.3 *Let M be the one-point blow-up of $\mathbb{CP}^2$. Then V is the exceptional divisor.*

Note that this V destabilizes slope stability in the sense of Ross-Thomas by a result of Panov and Ross [32].

Here is the outline of the proof of Theorem 5.3.2. Let $h \in C^\infty(M)$ satisfy $\mathrm{Ric}_g - \omega_g = i\partial\bar\partial h$. Suppose

$$\frac{\det(g_{i\bar j} + \varphi_{i\bar j})}{\det(g_{i\bar j})} = e^{-t\varphi + h}$$

has solutions only for $t \in [0, t_0)$, $t_0 < 1$. Then we have an MIS with support V. The following fact is due to Nadel based on earlier estimates by Siu and Tian.

Fact 5.3.4 Let $K \subset M - V$ be a compact subset of $M - V$. Then

$$\int_K \omega_{g_t}^m \to 0$$

as $t \to t_0$.

Fact 5.3.5

$$\mu_g(p) \in D^{\leq 0}(\xi) \iff (\mathrm{div}(v_\xi))(p) \geq 0$$

where

$$\mathrm{div}(v_\xi)(e^h \omega^m) = \mathcal{L}_{v_\xi}(e^h \omega^m).$$

Fact 5.3.6

$$\frac{t}{t-1} F(v_\xi) = \int_M \mathrm{div}(v_\xi)\omega_t^m.$$

By Fact 5.3.6 and our assumption $F(v_\xi) > 0$, we have for $t \in (\delta, t_0)$ with $t_0 < 1$

$$\int_M \text{div}(v_\xi)\, \omega_t^m = \frac{t}{t-1} F(v_\xi) < -C$$

with $C > 0$ independent of t.

We seek a contradiction by assuming $\mu_g(V) \subset D^{\leq 0}(\xi) = \{\text{div}(v_\xi) \geq 0\}$. Choose $\epsilon > 0$ small and put

$$W_\epsilon := \{p \in M \,|\, \text{div}(v_\xi)(p) \leq -\epsilon\}.$$

Then $W_\epsilon \subset M - V$ is compact. Apply Fact 5.3.4 to W_ϵ to get

$$\int_{W_\epsilon} \omega_{g_t}^m \to 0$$

as $t \to t_0$.

But then

$$-C \geq \int_M \text{div}(v_\xi)\omega_t^m = \int_{M-W_\epsilon} \text{div}(v_\xi)\omega_t^m + \int_{W_\epsilon} \text{div}(v_\xi)\omega_t^m$$
$$\geq -2\epsilon \text{vol}(M, g)$$

as $t \to t_0$, a contradiction! This completes the outline of the proof of Theorem 5.3.2.

Next, we turn to KRS-MIS. Let M be again a Fano manifold of dimension m. Let $\omega_g \in c_1(M)$ be a Kähler form and $v \in \mathfrak{h}_r(M)$ a holomorphic vector field in the reductive part $\mathfrak{h}_r(M)$ of $\mathfrak{h}(M)$.

Definition 5.3.7 The pair (g, v) is said to be a Kähler-Ricci soliton if

$$\text{Ric}(\omega_g) - \omega_g = \mathcal{L}_v(\omega_g).$$

(Hence $\text{Im}(v)$ is necessarily a Killing vector field.)

Start with an initial metric g^0 with $\omega_0 := \omega_{g^0} \in c_1(M)$.

$$\text{Ric}(\omega_0) - \omega_0 = i\partial\bar\partial h_0, \quad \int_M e^{h_0}\omega_0^m = \int_M \omega_0^m.$$

$$i_v \omega_0 = i\bar\partial\theta_{v,0}, \quad \int_M e^{\theta_{v,0}}\omega_0^m = \int_M \omega_0^m.$$

Consider for $t \in [0, 1]$

$$\det\left(g_{i\bar j}^0 + \varphi_{t i \bar j}\right) = \det\left(g_{i\bar j}^0\right)e^{h_0 - \theta_{v,0} - v\varphi_t - t\varphi_t}.$$

The solution for $t = 1$ gives the Kähler-Ricci soliton. Zhu [62] has shown that $t = 0$ always has a solution. The implicit function theorem shows for some $\epsilon > 0$, all $t \in [0, \epsilon)$ have a solution.

Suppose we only have solutions on $[0, t_\infty)$, $t_\infty < 1$.

Let $\theta_{v,g}$ satisfy

$$i_v \omega_g = i\bar{\partial}\theta_{v,g}, \quad \int_M e^{\theta_{v,g}} \omega_g^m = \int_M \omega_g^m.$$

Definition 5.3.8 Define $F_v : \mathfrak{h}(M) \to \mathbb{C}$ by

$$F_v(w) = \int_M w(h_g - \theta_{v,g}) e^{\theta_{v,g}} \omega_g^m.$$

Tian and Zhu [52] showed that this F_v is independent of g with $\omega_g \in c_1(M)$.

Theorem 5.3.9 (Tian-Zhu [52]) *There exists a unique $v \in \mathfrak{h}_r(M)$ such that*

$$F_v(w) = 0 \text{ for all } w \in \mathfrak{h}_r(M).$$

We take v to be the one chosen in the Theorem 5.3.9.

Theorem 5.3.10 ([21]) *Let K be the compact subgroup such that $\mathfrak{k} \otimes \mathbb{C} = \mathfrak{h}_r(M)$. Let v be the one chosen in the Theorem 5.3.9. Suppose there is no KRS. Then we get an MIS and its support V_s satisfies*

$$V_s \not\subset Z^+(\text{grad}'w) \text{ for } \forall v \in \mathfrak{h}_r(M).$$

Just as Nadel applied Theorem 5.3.1 to prove the existence of Kähler-Einstein metric on $\mathbb{CP}^1$, we can apply Theorem 5.3.10 to prove the existence of KRS on the one point blow-up of $\mathbb{CP}^2$.

Next we consider KRF-MIS. As mentioned in section 2, there are two approaches to KRF-MIS, one by Phong, Sesum and Sturm [37], and the other by Rubinstein [40]. Here, we consider the one considered by Rubinstein. So, one gets an MIS from the failure of convergence of normalized Kähler-Ricci flow:

$$\frac{\partial g}{\partial t} = -\text{Ric}(g) + g. \tag{5.20}$$

If we put $g_{ti\bar{j}} = g_{i\bar{j}} + \varphi_{ti\bar{j}}$ the Ricci flow is equivalent to

$$\frac{\partial \varphi_t}{\partial t} = \ln \frac{\det(g_{i\bar{j}} + \varphi_{ti\bar{j}})}{\det(g_{i\bar{j}})} + \varphi_t - h_0$$

$$\varphi_0 = c_0$$

Rubinstein modified Phong–Sesum–Sturm's MIS using the idea of Demailly–Kollár:

$$\varphi_t - \int_M \varphi_t \, \omega^m \longrightarrow \varphi_\infty \quad \text{almost psh}$$

as $t \to \infty$. Let V_γ be the MIS for $\psi = \gamma \varphi_\infty$, $\gamma \in (\frac{m}{m+1}, 1)$, defined by

$$\Gamma(U, \mathcal{I}(\psi)) = \left\{ f \in \mathcal{O}_M(U) \mid \int_U |f|^2 \, e^{-\psi} \, \omega_g^m < \infty \right\}.$$

This MIS satisfies

$$H^q(M, \mathcal{I}(\psi)) = 0 \quad \text{for} \quad \forall q > 0.$$

In general, it seems to be difficult to calculate (the support) of KRF-MIS. However, under some assumptions, it becomes computable. To explain it, we recall the following two results. Let M be a toric Fano manifold of dimension m, on which the algebraic torus $T_{\mathbb{C}} = (\mathbb{C}^*)^m$ acts. Let $T_{\mathbb{R}} = T^m$ be the real torus and $\mathfrak{t}_{\mathbb{R}}$ its Lie algebra. We further put $N_{\mathbb{R}} = J\mathfrak{t}_{\mathbb{R}}$. Let $W(M) = N(T_{\mathbb{C}})/T_{\mathbb{C}}$ be the Weyl group.

Theorem 5.3.11 (Wang-Zhu [56]) *There exists a Kähler-Ricci Soliton* $(g_{\mathrm{KRS}}, v_{\mathrm{KRS}})$.

Here we assume that K denotes a maximal compact subgroup of the reductive part of $\mathrm{Aut}(M)$ and $K_{v_{\mathrm{KRS}}}$ denotes the one-parameter subgroup of K generated by the imaginary part of v_{KRS}. Then,

Theorem 5.3.12 (Tian-Zhu [53]) *Let M be a (not necessarily toric) Fano manifold which admits a Kähler-Ricci soliton $(g_{\mathrm{KRS}}, v_{\mathrm{KRS}})$. Then, any solution g_t of (5.20) will converge to g_{KRS} in the sense of Cheeger-Gromov if the initial Kähler metric is $K_{v_{\mathrm{KRS}}}$-invariant.*

Combining Theorem 5.3.11 and Theorem 5.3.12, we find that the flow (5.20) always converges to a Kähler-Ricci soliton in the sense of Cheeger-Gromov on toric Fano manifolds. This fact suggests to us a possibility to understand the asymptotic behavior of g_t along (5.20) and to get some information about KRF-MIS from data of Kähler-Ricci solitons. In fact, the second author proved

Theorem 5.3.13 ([42]) *Suppose that the fixed point set $N_{\mathbb{R}}^{W(M)}$ of the Weyl group $W(M)$ on $N_{\mathbb{R}}$ is one dimensional. Let $\sigma_t = \exp(t v_{\mathrm{KRS}})$ be the one parameter group of transformations generated by v_{KRS}, $0 < \gamma < 1$ and ω a $T_{\mathbb{R}}$-invariant Kähler form in $c_1(M)$. Then the support of Rubinstein's KRF-MIS of exponent γ is equal to the support of the MIS of exponent γ obtained from the Kähler potentials of $\{(\sigma_t^{-1})^* \omega\}$.*

Note that the assumption of $N_{\mathbb{R}}^{W(M)}$ is constrained and it would be expected to be removed. Using the above theorem, the second author computed the support of KRF-MIS for various γ in some examples. For example, we can prove

Corollary 5.3.14 *Let M be the blow up of $\mathbb{CP}^2$ at p_1 and p_2. Let E_1 and E_2 be the exceptional divisors of the blow up, and E_0 be the proper transform of $\overline{p_1 p_2}$ of the line passing through p_1 and p_2. Then, the support of KRF-MIS on M of exponent γ is*

$$\begin{cases} \cup_{i=0}^2 E_i & \text{for } \gamma \in (\frac{1}{2}, 1), \\ E_0 & \text{for } \gamma \in (\frac{1}{3}, \frac{1}{2}). \end{cases}$$

It would be interesting to consider a relationship between destabilizing test configurations and the pair of the support of KRF-MIS and its exponent.

References

[1] T. Aubin : Equations du type de Monge-Ampère sur les variétés kählériennes compactes, C. R. Acad. Sci. Paris, **283**, 119–121 (1976)

[2] C.P. Boyer and K. Galicki : Sasakian geometry, (Oxford Mathematical Monographs., 2008).

[3] E. Calabi : Extremal Kähler metrics II, Differential geometry and complex analysis, (I. Chavel and H.M. Farkas eds.), 95–114, Springer-Verlag, Berline-Heidelberg-New York, (1985).

[4] H.D. Cao : Deformation of Kähler metrics to Kähler-Einstein metrics on compact Kähler manifolds, Invent. Math. 81 (1985) 359–372.

[5] I. Cheltsov and C. Shramov : Log canonical thresholds of smooth Fano threefolds (with an appendix by J.P. Demailly), Uspekhi Mat. Nauk 63 (2008), no. 5(383), 73–180; translation in Russian Math. Surveys 63 (2008), no. 5, 859–958.

[6] X.X. Chen and G. Tian : Geometry of Kähler metrics and foliations by holomorphic discs, Publ. Math. Inst. Hautes Études Sci. No. 107 (2008), 1–107. math.DG/0409433

[7] X.X. Chen and B. Wang : Remarks on Kähler Ricci flow, arXiv:0809.3963 (2008).

[8] K. Cho, A. Futaki and H. Ono : Uniqueness and examples of toric Sasaki-Einstein manifolds, Comm. Math. Phys., 277 (2008), 439–458, math.DG/0701122

[9] J.P. Demailly and J. Kollár : Semi-continuity of complex singularity exponents and Kähler-Einstein metrics on Fano orbifolds, Ann. Sci. École Norm. Sup. (4) 34, no.4 (2001) 525–556.

[10] S.K. Donaldson : Scalar curvature and stability of toric varieties, J. Differential Geometry, 62(2002), 289–349.

[11] S.K. Donaldson : Lower bounds on the Calabi functional, J. Differential Geometry, 70(2005), 453–472.

[12] S.K. Donaldson : Kähler geometry on toric manifolds, and some other manifolds with large symmetry, Handbook of geometric analysis. No. 1, 29–75, Adv. Lect. Math. (ALM), 7, Int. Press, Somerville, MA, 2008. arXiv:0803.0985 (2008).

[13] S.K. Donaldson : Remarks on gauge theory, complex geometry and four-manifold topology, in 'Fields Medallists Lectures' (Atiyah, Iagolnitzer eds.), World Scientific, 1997, 384–403.

[14] S.K. Donaldson and P.B. Kronheimer : The geometry of four manifolds, Oxford Mathematical Monographs, Claren Press, Oxford, 1990.

[15] A. Fujiki : Moduli space of polarized algebraic manifolds and Kähler metrics, Sugaku Expositions, 5(1992), 173–191.

[16] A. Futaki : An obstruction to the existence of Einstein Kähler metrics, Invent. Math. 73, 437–443 (1983).

[17] A. Futaki : On compact Kähler manifolds of constant scalar curvature, Proc. Japan Acad., Ser. A, 59, 401–402 (1983).

[18] A. Futaki : Kähler-Einstein metrics and integral invariants, Lecture Notes in Math., vol.1314, Springer-Verlag, Berline-Heidelberg-New York, (1988).

[19] A. Futaki : Stability, integral invariants and canonical Kähler metrics, Proc. 9-th Internat. Conf. on Differential Geometry and its Applications, 2004 Prague, (eds. J. Bures et al), 2005, 45–58, MATFYZPRESS, Prague.

[20] A. Futaki, H. Ono and G. Wang : Transverse Kähler geometry of Sasaki manifolds and toric Sasaki-Einstein manifolds, J. Differential Geom. 83 (2009), no. 3, 585–635.

[21] A. Futaki and Y. Sano : Multiplier ideal sheaves and integral invariants on toric Fano manifolds, arXiv:0711.0614 (2007).

[22] G. Heier : Convergence of the Kähler-Ricci flow and multiplier ideal sheaves on Del Pezzo surfaces, Michigan Math. J. 58 (2009), no. 2, 423–440. arXiv:0710.5725

[23] G. Heier : Existence of Kähler-Einstein metrics and multiplier ideal sheaves on Del Pezzo surfaces, Math. Z. 264 (2010), no. 4, 727–743. arXiv:0710.5724.

[24] J. Kollár : Singularities of pairs. (Algebraic Geometry–Santa Cruz 1995), Proc. Sympos. Pure Math., 62, part 1 (1995) 221–287.

[25] S. Kolodziej : The complex Monge-Ampère equation, Acta Math. 180, no.1 (1998) 69–117.

[26] S. Kolodziej : The Monge-Ampère equation on compact Kähler manifolds, Indiana Univ. Math. J. 52, no.3 (2003) 667–686.

[27] R. Lazarsfeld : Positivity in Algebraic Geometry II: Positivity for vector bundles, and multiplier ideals, (Springer-Verlag, 2004).

[28] T. Mabuchi : K-stability of constant scalar curvature polarization, arXiv:0812. 4093

[29] Y. Matsushima : Sur la structure du groupe d'homéomorphismes d'une certaine variété kaehlérienne, Nagoya Math. J., **11**, 145-150 (1957).

[30] A.M. Nadel : Multiplier ideal sheaves and Kähler-Einstein metrics of positive scalar curvature, Ann. of Math. (2) 132, no.3 (1990) 549–596.

[31] A.M. Nadel : Multiplier ideal sheaves and Futaki's invariant, Geometric Theory of Singular Phenomena in Partial Differential Equations (Cortona, 1995), Sympos. Math., XXXVIII, Cambridge Univ. Press, Cambridge, 1998. (1995) 7–16.

[32] D. Panov and J. Ross : Slope Stability and Exceptional Divisors of High Genus, Math. Ann. 343 (2009), no. 1, 79–101. arXiv:0710.4078v1 [math.AG].

[33] S.T. Paul and G. Tian : CM Stability and the Generalized Futaki Invariant I., math.AG/0605278, 2006.

[34] S.T. Paul and G. Tian : CM Stability and the Generalized Futaki Invariant II. (To appear in Asterisque), math.AG/0606505, 2006.

[35] D.H. Phong and J. Sturm : On stability and the convergence of the Kähler-Ricci flow, J. Differential Geom. 72 (2006), no. 1, 149–168.

[36] D.H. Phong and J. Sturm : Lectures on stability and constant scalar curvature, Handbook of geometric analysis, No. 3, 357–436, Adv. Lect. Math. (ALM), 14, Int. Press, Somerville, MA, 2010. arXiv:0801.4179.

[37] D.H. Phong, N. Sesum and J. Sturm : Multiplier ideal sheaves and the Kähler-Ricci flow, Comm. Anal. Geom. 15, no. 3 (2007), 613–632.

[38] J. Ross and R.P. Thomas : An obstruction to the existence of constant scalar curvature Kähler metrics, J. Differential Geom. 72 (2006), no. 3, 429–466.

[39] Y.A. Rubinstein : The Ricci iteration and its applications, C. R. Acad. Sci. Paris, Ser. I, 345 (2007), 445-448, arXiv:0706.2777.

[40] Y.A. Rubinstein : On the construction of Nadel multiplier ideal sheaves and the limiting behavior of the Ricci flow, Trans. Amer. Math. Soc. 361 (2009), no. 11, 5839–5850. arXive:math/0708.1590.

[41] Y.A. Rubinstein : Some discretizations of geometric evolution equations and the Ricci iteration on the space of Kähler metrics, Adv. Math. 218 (2008), no. 5, 1526–1565. arXive:math/0709.0990.

[42] Y. Sano : Multiplier ideal sheaves and the Kähler-Ricci flow on toric Fano manifolds with large symmetry, preprint, arXiv:0811.1455.

[43] N. Sesum and G. Tian : Bounding scalar curvature and diameter along the Kähler-Ricci flow (after Perelman), J. Inst. Math. Jussieu 7 (2008), no. 3, 575–587.

[44] Y-T. Siu : The existence of Kähler-Einstein metrics on manifolds with positive anticanonical line bundle and a suitable finite symmetry group, Annals of Math. 127 (1988) 585–627.

[45] Y-T. Siu : Dynamical multiplier ideal sheaves and the construction of rational curves in Fano manifolds, Complex analysis and digital geometry, 323–360, Acta Univ. Upsaliensis Skr. Uppsala Univ. C Organ. Hist., 86, Uppsala Universitet, Uppsala, 2009. arXiv:0902.2809.

[46] J. Song : The α-invariant on toric Fano manifolds, Amer. J. Math. 127, No.6 (2005) 1247–1259.

[47] J. Stoppa : K-stability of constant scalar curvature Kähler manifolds, Adv. Math. 221 (2009), no. 4, 1397–1408. arXiv:0803.4095.

[48] G. Székelyhidi : The Kähler-Ricci flow and K-polystability, Amer. J. Math. 132 (2010), no. 4, 1077–1090. arXiv:0803.1613.

[49] G. Tian : On Kähler-Einstein metrics on certain Kähler manifolds with $C_1(M) > 0$, Invent. Math. 89, no.2 (1987) 225–246.

[50] G. Tian, Kähler-Einstein metrics with positive scalar curvature, Invent. Math. 130, no.1 (1997) 1–37.

[51] G. Tian and S.T. Yau, Kähler-Einstein metrics on complex surfaces with $C_1(M)$ positive, Comm. Math. Phys. 112 (1987) 175–203.

[52] G. Tian and X. Zhu : A new holomorphic invariant and uniqueness of Kähler-Ricci solitons, Comment. Math. Helv 77, No.2 (2002) 297–325.

[53] G. Tian and X. Zhu : Convergence of Kähler-Ricci flow, J. Amer. Math. Soc., 20, No.3 (2007), 675–699.

[54] V. Tosatti : Kähler-Ricci flow on stable Fano manifolds, J. Reine Angew. Math. 640 (2010), 67–84. arXiv:0810.1895.

[55] X.-W. Wang : Moment maps, Futaki invariant and stability of projective manifolds, Comm. Anal. Geom. 12 (2004), no. 5, 1009–1037.

[56] X.J. Wang and X. Zhu : Kähler-Ricci solitons on toric manifolds with positive first Chern class, Adv. Math. 188, No.1 (2004) 87–103.

[57] B. Weinkove, A complex Frobenius theorem, multiplier ideal sheaves and Hermitian-Einstein metrics on stable bundles, Trans. Amer. Math. Soc. 359, no.4 (2007) 1577–1592.

[58] S.-T.Yau : On the Ricci curvature of a compact Kähler manifold and the complex Monge-Ampère equation I, Comm. Pure Appl. Math. 31(1978), 339–441.

[59] S.-T. Yau : Open problems in Geometry, Proc. Symp. Pure Math. 54 (1993) 1–28.

[60] R. Ye : The logarithmic Sobolev inequality along the Ricci flow, arXiv:0707.2424 (2007).

[61] O.S. Zhang : A uniform Sobolev inequality under Ricci flow, arXiv:0706.1594 (2007).

[62] X. Zhu, Kähler-Ricci soliton type equations on compact complex manifolds with $C_1(M) > 0$, J. Geom. Anal., 10 (2000), 759–774.

Authors' addresses:

Akito Futaki
Department of Mathematics,
Tokyo Institute of Technology, 2-12-1,
O-okayama, Meguro, Tokyo 152-8551,
Japan
futaki@math.titech.ac.jp

Yuji Sano
Department of Mathematics,
Kyushu University,
6-10-1, Hakozaki, Higashiku,
Fukuoka-city, Fukuoka 812-8581
Japan
sano@math.kyushu-u.ac.jp

6

Multisymplectic formalism and the covariant phase space

FRÉDÉRIC HÉLEIN

In most attempts to build the mathematical foundations of Quantum Field Theory (QFT), two classical ways have been explored. The first one is often referred to as the *Feynman integral* or *functional integral method*. It is a generalization to fields of the path integral method of quantum mechanics and is heuristically based on computing integrals over the infinite dimensional set of all possible fields φ by using a kind of 'measure' – which should behave like the Lebesgue measure on the set of all possible fields φ – times $e^{i\mathcal{L}(\varphi)/\hbar}$, where $\mathcal{L}$ is a Lagrangian functional (but attempts to define this 'measure' failed in most cases). The second one is referred to as the *canonical quantization method* and is based on the Hamiltonian formulation of the dynamics of classical fields, by following general axioms which were first proposed by Dirac and later refined. The Feynman approach has the advantage of being manifestly relativistic, i.e. it does not require the choice of a particular system of space-time coordinate, since the main ingredient is $\mathcal{L}(\varphi)$, which is an integral over all space-time. By contrast, the canonical approach, at least its classical formulation, seems to be based on the choice of a particular time coordinate which is needed to define the Hamiltonian function through an infinite dimensional Legendre transform.

However there are alternative formulations of the Hamiltonian structure of the dynamics of classical fields, which could be used as a starting point of a *covariant canonical quantization*[1]. We shall see two of them in this text: the *covariant phase space* and the *multisymplectic formalism*. The covariant phase space is based on the observation that the set of classical solutions to a variational problem, i.e. of critical points of some action functional $\mathcal{L}(\varphi)$, has an intrinsic canonical symplectic (or presymplectic) structure. The multisymplectic formalism is a generalization of the standard symplectic formalism, where

---

[1] The word '*covariant*' refers here to a construction which does not rely on the choice of a particular system of coordinates on space-time and hence which respects the basic principles of Relativity.

the *time* of classical mechanics is replaced by *space-time*: for instance if we start from a Lagrangian action $\mathcal{L}(\varphi) = \int_{\mathcal{X}} L(x, \varphi, d\varphi)$ we do not perform a Legendre transform with respect to a chosen time coordinate, but with respect to all space-time coordinates.

We expect that, roughly speaking, both the Feynman and the canonical approach should lead to equivalent theories. However if this fact is true, it should not be trivial for several reasons. A first obvious remark is that both theories are only heuristic and have no mathematical foundations, except in very simplified situations. A superficial difference between both approaches is the fact that one is based on the Lagrangian, the other one the Hamiltonian function. Moreover these two approaches answer different questions: Feynman's offers a short and intuitive way to compute the quantities which can be measured in interaction processes between particles (although one cannot avoid the difficult step of regularizing and renormalizing the computed quantities). For the same task, the canonical approach seems to be more complicated; however it proposes a scheme to build mathematical objects (a complex Hilbert space of physical *states* and an algebra of self-adjoint operators acting on it, corresponding to *observable quantities*), the construction of which requires more effort by using the Feynman integral. But a deep difference between the methods is that the Feynman integral is a construction *off shell*, i.e. on the set of *all* possible fields, even those which are not solutions of the classical dynamical equations, whereas in many cases the canonical approach is a construction *on shell*, i.e. on the set of fields which are solutions of the dynamical equations (in particular in the covariant phase space method).

In this paper we shall present briefly the multisymplectic formalism and the covariant phase space and show the strong relation between both theories. To my knowledge this relation was discovered by J. Kijowski and W. Szczyrba in 1976 [25], but their beautiful paper seems to have been ignored in the literature. We have included some historical comments. We shall conclude by presenting the geometric quantization scheme for linear field equations (i.e. *free fields* in the terminology of physicists) in the framework of multisymplectic geometry. The goal is to show how canonical quantization can be performed in a covariant way.

6.1 The multisymplectic formalism

6.1.1 Maps between vector spaces

We start with a simple variational problem: let X and Y be two vector space of dimension n and k respectively and assume that X is oriented, let U be an open subset of X and consider the set $\mathcal{C}^\infty(U, Y)$ of smooth maps $\mathbf{u}$ from U to Y. Let $L : U \times Y \times \text{End}(X, Y) \longrightarrow \mathbb{R}$ be a Lagrangian density and consider the action functional on $\mathcal{C}^\infty(U, Y)$ defined by:

$$\mathcal{L}[\mathbf{u}] = \int_U L(x, \mathbf{u}(x), d\mathbf{u}_x)\beta,$$

where β is a volume n-form on U. We use coordinates $(x^1, \ldots, x^n)$ on U s.t. $\beta = dx^1 \wedge \cdots \wedge dx^n$, coordinates $(y^1, \ldots, y^k)$ on Y and v_μ^i on End(X, Y). Then the critical points of $\mathcal{L}$ satisfy the Euler–Lagrange system of equations

$$\frac{\partial}{\partial x^\mu}\left(\frac{\partial L}{\partial v_\mu^i}(x, \mathbf{u}(x), d\mathbf{u}_x)\right) = \frac{\partial L}{\partial y^i}(x, \mathbf{u}(x), d\mathbf{u}_x), \quad \forall i \text{ s.t. } 1 \leq i \leq k. \quad (6.1)$$

We assume that the map

$$U \times Y \times \text{End}(X, Y) \longrightarrow U \times Y \times \text{End}(Y^*, X^*)$$
$$(x, y, v) \longmapsto \left(x, y, \tfrac{\partial L}{\partial v}(x, y, v)\right),$$

is a diffeomorphism: this is the analogue of the *Legendre hypothesis* used in Mechanics. We denote by $p^* = (p_i^\mu)_{\mu,i}$ the coordinates on End(Y*, X*) and we define the Hamiltonian function H on $U \times Y \times \text{End}(Y^*, X^*)$ by

$$H(x, y, p^*) := p_i^\mu v_\mu^i - L(x, y, v),$$

where we assume implicitly that $v = (v_\mu^i)_{i,\mu}$ is the unique solution of $\frac{\partial L}{\partial v}(x, y, v) = p^*$. Then v_μ^i is actually equal to $\frac{\partial H}{\partial p_i^\mu}$. Moreover to any map $\mathbf{u} : U \longrightarrow Y$ we associate the map $\mathbf{p}^* : U \longrightarrow \text{End}(Y^*, X^*)$ s.t. $\mathbf{p}^*(x) := \frac{\partial L}{\partial v}(x, \mathbf{u}(x), d\mathbf{u}_x), \forall x \in U$. Then we can show [45] that $\mathbf{u}$ is a solution of (6.1) iff $(\mathbf{u}, \mathbf{p}^*)$ is a solution of the generalized Hamilton system:

$$\begin{cases} \dfrac{\partial \mathbf{u}^i}{\partial x^\mu}(x) = \dfrac{\partial H}{\partial p_i^\mu}(x, \mathbf{u}(x), \mathbf{p}^*(x)) \\[2mm] \dfrac{\partial \mathbf{p}_i^\mu}{\partial x^\mu}(x) = -\dfrac{\partial H}{\partial y^i}(x, \mathbf{u}(x), \mathbf{p}^*(x)). \end{cases} \quad (6.2)$$

System (6.2) can be translated as a geometric condition [23] on the graph

$$\Gamma^* := G(\mathbf{u}, \mathbf{p}^*) := \{(x, \mathbf{u}(x), \mathbf{p}^*(x))| \, x \in U\} \subset U \times Y \times \text{End}(Y^*, X^*).$$

Indeed consider a family of n vector fields $X_1, \ldots, X_n : U \longrightarrow U \times Y \times \text{End}(Y^*, X^*)$ s.t. for any $x \in U$, $(X_1(x), \ldots, X_n(x))$ is a basis of the tangent plane to $G(\mathbf{u}, \mathbf{p}^*)$ at $(x, \mathbf{u}(x), \mathbf{p}^*(x))$. Set $\beta_\mu := \frac{\partial}{\partial x^\mu} \lrcorner \beta$. Then (6.2) is equivalent to the condition that $\forall \xi \in X \times Y \times \text{End}(Y^*, X^*)$,

$$dp_i^\mu \wedge dy^i \wedge \beta_\mu(\xi, X_1, \ldots, X_n) = dH(\xi)\beta(X_1, \ldots, X_n). \quad (6.3)$$

In fact this can be easily checked by choosing $X_\mu := \frac{\partial}{\partial x^\mu} + \frac{\partial u^i}{\partial x^\mu}\frac{\partial}{\partial y^i} + \frac{\partial p_i^\mu}{\partial x^\mu}\frac{\partial}{\partial p_i^\mu}$.
More concisely we can introduce the n-multivector field $X := X_1 \wedge \cdots \wedge X_n$
(so that $X(x) \in \Lambda^n T_{(x,\mathbf{u}(x),\mathbf{p}^*(x))}\Gamma^*, \forall x \in U$). Then Equation (6.3) reads:

$$\forall \xi \in X \times Y \times \mathrm{End}(Y^*, X^*), \quad dp_i^\mu \wedge dy^i \wedge \beta_\mu(\xi, X) = dH(\xi)\beta(X). \quad (6.4)$$

Equation (6.3) can be completed with the independence condition

$$\beta|_{\Gamma^*} \neq 0, \quad (6.5)$$

where, if $j_{\Gamma^*} : \Gamma^* \longrightarrow U \times Y \times \mathrm{End}(Y^*, X^*)$ denotes the inclusion map,
$\beta|_{\Gamma^*} := j_{\Gamma^*}^* \beta$. This condition guarantees that locally Γ^* is the graph of some
map $(\mathbf{u}, \mathbf{p}^*)$ over the 'space-time' X.

We will see now that the independence condition (6.5) can be further incorporated in a dynamical condition analogous to (6.3) by adding to the variables
(x, y, p^*) a variable e dual to β. We define $M := U \times Y \times \mathbb{R} \times \mathrm{End}(Y^*, X^*)$
with coordinates $(x, y, e, p^*) = (x^\mu, y^i, e, p_i^\mu)$ and the $(n+1)$-form

$$\omega := de \wedge \beta + dp_i^\mu \wedge dy^i \wedge \beta_\mu. \quad (6.6)$$

We define $\mathcal{H} : M \longrightarrow \mathbb{R}$ by $\mathcal{H}(x, y, e, p^*) := e + H(x, y, p^*)$. Then to any
oriented n-dimensional submanifold $\Gamma^* = G(\mathbf{u}, \mathbf{p}^*)$ we associate the oriented
n-dimensional submanifold $\Gamma := \{(x, \mathbf{u}(x), \mathbf{e}(x), \mathbf{p}^*(x)) | x \in U\}$ of M, where
$\mathbf{e}$ is such that $\mathbf{e}(x) + H(x, \mathbf{u}(x), \mathbf{p}^*(x)) = h, \forall x \in X$, for some real constant[2]
h. Then Γ^* is a solution of (6.3) and (6.5) iff Γ is a solution of:

$$\forall \xi \in \mathcal{C}^\infty(M, TM), \quad \omega(\xi, X) = d\mathcal{H}(\xi)\beta(X), \quad (6.7)$$

where $\mathcal{C}^\infty(M, TM)$ denotes the set of sections of TM over M, which can be
identified with $\mathcal{C}^\infty(M, X \times Y \times \mathbb{R} \times \mathrm{End}(Y^*, X^*))$. Note that conversely it is
easy to check that any connected solution Γ of (6.7) is contained in a level set of
$\mathcal{H}$. We call a **Hamiltonian n-curve** any solution Γ of (6.7). The $(n+1)$-form
ω is an example of a multisymplectic form and the pair (M, ω) is called a
multisymplectic manifold. Using the notation $\xi \lrcorner \omega$ for the interior product of
the vector ξ with the $(n+1)$-form ω, we set:

Definition 6.1.1 Let $\mathcal{M}$ be a smooth manifold. A **multisymplectic $(n+1)$-
form** ω on $\mathcal{M}$ is a $(n+1)$-form which is **closed** (i.e. $d\omega = 0$) and which is
non degenerate (i.e. $\forall M \in \mathcal{M}, \forall \xi \in T_M\mathcal{M}, \xi \lrcorner \omega = 0 \implies \xi = 0$).

---

[2] W.l.g. we can assume that the constant h is zero, so that Γ is included in $\mathcal{H}^{-1}(0)$.

6.1.2 Higher order problems

The preceding can be extended to an action on maps $\mathbf{u} : U \longrightarrow Y$ of the form $\mathcal{L}[\mathbf{u}] := \int_U L(x, j^r\mathbf{u}(x))\beta$, where $j^r\mathbf{u}$ denotes the r-th order jet of $\mathbf{u}$ (i.e. all partial derivatives of $\mathbf{u}$ of order less than or equal to r). We denote by $J^r(U, Y)$ the r-th order jet space of maps from U to Y and we use the coordinates $x = (x^\mu)_\mu$ and $v = \left(v^i_{\mu_1\cdots\mu_a}\right)_{i,\mu_1\cdots\mu_a}$ (for $1 \le i \le k, 0 \le a \le r$ and $1 \le \mu_b \le n$) on $J^r(U, Y)$ s.t.

$$v^i_{\mu_1\cdots\mu_a}\left(j^r\mathbf{u}(x)\right) = \frac{\partial^a\mathbf{u}^i}{\partial x^{\mu_1}\cdots\partial x^{\mu_a}}(x).$$

It is convenient to introduce the multi-index notation $M = \mu_1\cdots\mu_a$, where $a \in \mathbb{N}$ and $\forall b \in [\![1, a]\!], 1 \le \mu_b \le n$ and to set $|M| = a$. Then for $|M| = r$ we denote

$$\pi^M_i(x, v) := \frac{\partial L}{\partial v^i_M}(x, v). \tag{6.8}$$

The analogue of the Legendre hypothesis consists here in supposing that the map $(v^i_M)_{i,M;|M|=r} \longmapsto (p^M_i)_{i,M;|M|=r}$ defined by (6.8) is one to one. Next we define the vector space M with coordinates:

$$
\begin{array}{cccccc}
(x, v) & : & x^\mu & v^i & v^i_\mu & \cdots & v^i_{\mu_1\cdots\mu_{r-1}} \\
p = (e, p^*) & : & & e & p^\mu_i & \cdots & p^{\mu_1\cdots\mu_{r-1}}_i & p^{\mu_1\cdots\mu_r}_i
\end{array}
$$

for $1 \le \mu, \mu_b \le n$ and $1 \le i \le k$. Clearly M contains $J^{r-1}(U, Y)$ as a vector subspace. We also define recursively, for $|M| \le r - 1$,

$$\pi^M_i(x, v) := \frac{\partial L}{\partial v^i_M}(x, v) - D_\mu \pi^{M\mu}_i(x, v_M),$$

where

$$D_\mu := \frac{\partial}{\partial x^\mu} + v^i_{M\mu}\frac{\partial}{\partial v^i_M}.$$

Then we define a Hamiltonian function on M:

$$H(x, v, p^*) := p^\mu_i v^i_\mu + \cdots + p^{\mu_1\cdots\mu_r}_i v^i_{\mu_1\cdots\mu_r} - L(x, v),$$

where we assume implicitly that, for $|M| = r$, $v^i_M = v^i_{\mu_1\cdots\mu_r}$ is the solution of $p^M_i = \pi^M_i(x, v)$, $\forall M$ s.t. $|M| = r$, and we set $p^M_i := \pi^M_i(x, v)$, $\forall M$ s.t. $|M| < r$. To any map $\mathbf{u}$ from U to Y we associate the map $\mathbf{p}^*$ which is the image of $j^r\mathbf{u}$ by the maps π^M_i. Then $\mathbf{u}$ is a critical point of $\mathcal{L}$ iff $(j^{r-1}\mathbf{u}, \mathbf{p}^*)$ is a solution of the generalized Hamilton equations [8]

$$
\begin{cases}
\dfrac{\partial \mathbf{u}^i_{\mu_1 \cdots \mu_a}}{\partial x^\mu}(x) = \dfrac{\partial H}{\partial p_i^{\mu_1 \cdots \mu_a \mu}}(x, \mathbf{u}(x), \mathbf{p}^*(x)) & \text{for } 0 \le a \le r - 1 \\[3mm]
\dfrac{\partial \mathbf{p}_i^{\mu_1 \cdots \mu_a \mu}}{\partial x^\mu}(x) = -\dfrac{\partial H}{\partial v^i_{\mu_1 \cdots \mu_a}}(x, \mathbf{u}(x), \mathbf{p}^*(x)) & \text{for } 0 \le a \le r - 1,
\end{cases}
\tag{6.9}
$$

Alternatively we can consider the map $x \longmapsto (x, j^{r-1}\mathbf{u}(x), \mathbf{e}(x), \mathbf{p}^*(x))$, where $\mathbf{e}$ may be chosen so that $\mathbf{e}(x) + H(x, j^{r-1}\mathbf{u}(x), \mathbf{p}^*(x)) = 0$, $\forall x$ and we can write (6.9) in a way similar to (6.7) by using the Hamiltonian function

$$
\mathcal{H}(x, v, p) = e + H(x, v, p^*).
$$

and the multisymplectic form

$$
\omega := de \wedge \beta + dp_i^\mu \wedge dv^i \wedge \beta_\mu + \cdots + dp_i^{\mu_1 \cdots \mu_r} \wedge dv^i_{\mu_1 \cdots \mu_{r-1}} \wedge \beta_{\mu_r}.
$$

An intrinsic geometrical multisymplectic formulation of these equations has been derived recently by L. Vitagliano [43].

6.1.3 More general multisymplectic manifolds

Assume that we start with an action $\mathcal{L}$ which is an integral of a Lagrangian density which depends on the first order derivatives of the field. This may be for instance a variational problem on maps $\mathbf{u}$ between two manifolds $\mathcal{X}$ and $\mathcal{Y}$ with a functional $\mathcal{L}[\mathbf{u}] := \int_{\mathcal{X}} L(x, \mathbf{u}(x), d\mathbf{u}_x)\beta$ or a variational problem on sections of a fiber bundle $\pi : \mathcal{Z} \longrightarrow \mathcal{X}$. Then a natural multisymplectic manifold is the vector bundle $\Lambda^n T^*(\mathcal{X} \times \mathcal{Y})$ in the first case or $\Lambda^n T^* \mathcal{Z}$ in the second case. Both manifolds are indeed endowed with a canonical $(n + 1)$-form ω which is the straightforward analogue of the canonical symplectic form on any cotangent bundle [6, 24, 19]. We may call this manifold the *universal multisymplectic manifold* associated with the Lagrangian problem. Although this construction seems to be similar to the symplectic one for Hamiltonian mechanics, it is different because, say for maps between two manifolds $\mathcal{X}$ and $\mathcal{Y}$ of dimensions n and k respectively and a first order variational problem, on the one hand the Lagrangian density depends on $n + k + nk$ variables (in other words the analogue of the product of the time real line and of the tangent bundle in mechanics has dimension $n + k + nk$), whereas on the other hand the analogue of the cotangent bundle is $\Lambda^n T^*(\mathcal{X} \times \mathcal{Y})$ and has dimension $n + k + \frac{(n+k)!}{n!k!}$. This means that we have much more choices in the Legendre transform, which is not a map in general but a correspondence, as soon as $n \ge 2$ and $k \ge 2$.

This is why it is possible and often simpler to impose extra constraints in the Legendre transform, which means that we replace the universal multisymplectic manifold $\Lambda^n T^* \mathcal{Z}$ (whatever $\mathcal{Z}$ is: a Cartesian product $\mathcal{X} \times \mathcal{Y}$ or the total space of a bundle) by some submanifold of it. Most authors prefer to use the *affine multisymplectic submanifold* $\Lambda_2^n T^* \mathcal{Z}$: if $\mathcal{Z} = \mathcal{X} \times \mathcal{Y}$, it is the subbundle of $\Lambda^n T^* (\mathcal{X} \times \mathcal{Y})$ over $\mathcal{X} \times \mathcal{Y}$, the fiber over the point $(x, y) \in \mathcal{X} \times \mathcal{Y}$ of which is the subspace of n-forms $p \in \Lambda^n T^*_{(x,y)}(\mathcal{X} \times \mathcal{Y})$ s.t. $\forall \eta_1, \eta_2 \in T_y \mathcal{Y}$, $(0, \eta_1) \wedge (0, \eta_2) \lrcorner p = 0$. If $\mathcal{Z}$ is a fiber bundle over $\mathcal{X}$, $\Lambda_2^n T^* \mathcal{Z}$, which is the subbundle over $\mathcal{Z}$, the fiber over $z \in \mathcal{Z}$ of which is the subspace of n-forms $p \in \Lambda^n T^*_z \mathcal{Z}$ s.t. for any pair of *vertical* vectors $\eta_1, \eta_2 \in T_z \mathcal{Z}$, $\eta_1 \wedge \eta_2 \lrcorner p = 0$ (by 'vertical' we mean that η_1 and η_2 are maps to $0 \in T_{\pi(z)} \mathcal{X}$ by the differential of π). In both cases an n-form reads $p = e\beta + p_i^\mu dy^i \wedge \beta_\mu$ in local coordinates and the latter theory is actually the right generalization of (6.6). This theory is usually referred to as the *De Donder–Weyl theory* although it was discovered by V. Volterra (see §6.1.8). Note that $\Lambda_2^n T^* \mathcal{Z}$ can alternatively be defined as being the affine dual of the first jet bundle of sections of $\mathcal{Z}$ over $\mathcal{X}$ [14].

6.1.4 Premultisymplectic manifolds

A variant consists in manifolds equipped with a closed $(n + 1)$-form but without assuming a non-degeneracy condition, as for instance in [25]:

Definition 6.1.2 We call a triple $(\mathcal{M}, \omega, \beta)$ an n-**phase space** if $\mathcal{M}$ is a manifold, ω is a closed $(n + 1)$-form, called a **premultisymplectic form** and β is a non vanishing n-form.

Examples of premultisymplectic manifolds can be built easily by starting from a multisymplectic manifold $(\mathcal{M}, \omega)$ with a Hamiltonian function $\mathcal{H}$ on it which has no critical points (as for instance $\mathcal{H}(x, y, e, p^*) = e + H(x, y, p^*)$ for the previous theory). Then we let η be a vector field on $\mathcal{M}$ s.t. $d\mathcal{H}(\eta) = 1$ everywhere and we set $\beta := \eta \lrcorner \omega$. For any $h \in \mathbb{R}$ the level set $\mathcal{M}^h := \mathcal{H}^{-1}(h)$ is a submanifold. Then $(\mathcal{M}^h, \omega|_{\mathcal{M}^h}, \beta|_{\mathcal{M}^h})$ is a premultisymplectic manifold [18]. In particular $\omega|_{\mathcal{M}^h}$ is obviously closed but may be degenerate in general: indeed if Γ is a Hamiltonian n-curve contained in $\mathcal{M}^h$ then any vector tangent to Γ is in the kernel of $\xi \longmapsto \xi \lrcorner \omega$. In fact an n-phase space $(\mathcal{M}, \omega)$ carries an intrinsic dynamical structure: we say that *an n-dimensional submanifold Γ of $\mathcal{M}$ is a Hamiltonian n-curve if:*

$$\forall v \in \mathcal{C}^\infty(\mathcal{M}, T_\mathcal{M} \mathcal{M}), \quad (v \lrcorner \omega)|_\Gamma = 0 \quad \text{and } \beta|_\Gamma \neq 0. \tag{6.10}$$

This definition is motivated by the fact that (if Γ is connected) Γ is a solution of (6.7) iff there exists some $h \in \mathbb{R}$ s.t. Γ is contained in $\mathcal{M}^h$ and Γ is a Hamiltonian n-curve in the n-phase space $(\mathcal{M}^h, \omega|_{\mathcal{M}^h}, \beta|_{\mathcal{M}^h})$ (see [18]). However there are examples of premultisymplectic manifolds which do no arise from this construction as for instance the example in [20, 35] obtained by starting from the Palatini formulation of gravity.

6.1.5 Action principle

We assume here that we are working in a premultisymplectic manifold $(\mathcal{M}, \omega, \beta)$ and that the form ω is exact, i.e. is of the form $\omega = d\theta$, where θ is an n-form on $\mathcal{M}$. This is true e.g. in a submanifold of $\Lambda^n T^* \mathcal{Z}$, where ω is precisely defined as the differential of a canonical 'Poincaré–Cartan' form θ. To any oriented n-dimensional submanifold Γ we associate the action

$$\mathcal{A}[\Gamma] := \int_\Gamma \theta. \tag{6.11}$$

One can then show that any n-dimensional submanifold Γ on which β does not vanish is a critical point of $\mathcal{A}$ iff it is a Hamiltonian n-curve, i.e. a solution of (6.10) (see [18]). Actually if Γ is the image of a given configuration by some Legendre transform, then $\mathcal{A}[\Gamma]$ coincides with the Lagrangian action of the configuration we started with [19]. Note that in the case where ω is not exact one could define a similar action on a homology class of n-dimensional submanifolds by replacing $\int_\Gamma$ by $\int_\Delta \omega$, where Δ is a $(n+1)$-chain connecting Γ with a particular n-dimensional submanifold which generates the homology class.

6.1.6 Observable functionals

An observable functional is a functional on the 'space' of all solutions: this notion will be central in the next section concerning the covariant phase space. A particular class of such functionals arise in the context of multisymplectic manifolds or premultisymplectic manifolds as follows. In the following we denote by $\mathcal{F}$ the set of n-dimensional oriented submanifolds (*fields*) of $\mathcal{M}$ and by $\mathcal{E}$ the subset of $\mathcal{F}$ composed of Hamiltonian n-curves.

In a multisymplectic manifold $(\mathcal{M}, \omega)$

We define an **infinitesimal symplectomorphism of** $(\mathcal{M}, \omega)$ to be a vector field $\xi \in \mathcal{C}^\infty(\mathcal{M}, T\mathcal{M})$ s.t. $L_\xi \omega = 0$ (i.e. the Lie derivative of ω by ξ vanishes). Note that since ω is closed, this relation is equivalent ot $d(\xi \lrcorner \omega) = 0$. An important case occurs when $\xi \lrcorner \omega$ is exact: then there exists an $(n-1)$-form F s.t.

$$dF + \xi \lrcorner \omega = 0. \tag{6.12}$$

Any $(n-1)$-form F on $\mathcal{M}$ s.t. there exists a vector field ξ satisfying (6.12) is called an **observable** $(n-1)$**-form**. In the case where $n = 1$ then F is a function and in fact any function on $\mathcal{M}$ is an observable 0-form because the symplectic form is non degenerate. However if $n \geq 2$ then an arbitrary $(n-1)$-form on $\mathcal{M}$ is *not observable in general*, but if it is so then the vector field ξ s.t. (6.12) holds is *unique*: we shall denote it by ξ_F. Observable $(n-1)$-forms can be integrated over hypersurfaces in an n-curve to produce observable functionals. For that purpose, given some Hamiltonian function $\mathcal{H}$ on $\mathcal{M}$ we define a *slice* Σ to be codimension one submanifold of $\mathcal{M}$ s.t. for any Hamiltonian n-curve Γ the intersection of Σ with Γ is transverse. We also assume that Σ is co-oriented, which means that $\forall M \in \Sigma$ the 1-dimensional quotient space $T_M\mathcal{M}/T_M\Gamma$ is oriented. Then we can endow $\Sigma \cap \Gamma$ with an orientation and define

$$\int_\Sigma F : \mathcal{F} \longrightarrow \mathbb{R}$$
$$\Gamma \longmapsto \int_{\Sigma \cap \Gamma} F$$

Then one can recover two important notions in the semi-classical theory of fields. First one can define a bracket between observable $(n-1)$-forms F and G by the formula

$$\{F, G\} := \xi_F \wedge \xi_G \lrcorner\, \omega = \xi_F \lrcorner\, dG = -\xi_G \lrcorner\, dF.$$

Obviously $\{F, G\}$ is also an $(n-1)$-form. Moreover one can prove that it is also observable and that $\xi_{\{F,G\}} = [\xi_F, \xi_G]$ [23, 20]. Then the set of observable $(n-1)$-forms equipped with this 'Poisson bracket' becomes almost a Lie algebra (it satisfies the antisymmetry relation $\{F, G\} + \{G, F\} = 0$, but not the Jacobi identity; we have instead $\{\{G, H\}, F\} + \{\{H, F\}, G\} + \{\{F, G\}, H\} = d(\xi_F \wedge \xi_G \wedge \xi_H \lrcorner\, \omega)$, which, in the case where $n = 2$ can be understood as a Lie 2-algebra structure [1]). However we can define the bracket

$$\left\{ \int_\Sigma F, \int_\Sigma G \right\} := \int_\Sigma \{F, G\} \qquad (6.13)$$

which coincides with the Poisson bracket on functionals on fields used by physicists. We will also meet an interpretation of this bracket in the next Section.

A second important notion is the relation between observable forms and the dynamics. Indeed if Γ is a Hamiltonian n-curve and if F is an observable $(n-1)$-form then one can use the dynamical equation (6.7) with the vector field ξ_F. It gives us, $\forall M \in \Gamma$, $\forall X \in \Lambda^n T_M\Gamma$,

$$dF(X) = -\omega(\xi_F, X) = -d\mathcal{H}(\xi_F)\beta(X). \qquad (6.14)$$

Hence we see that *if $d\mathcal{H}(\xi_F)$ vanishes*, then $dF|_\Gamma$ vanishes. This implies by using Stokes theorem that the restriction of the functional $\int_\Sigma F$ to the set $\mathcal{E}$ of Hamiltonian n-curves does not depend on Σ but on its homology class. For that reason we say that **an observable $(n-1)$-form F is dynamical if** $d\mathcal{H}(\xi_F) = 0$.

In a premultisymplectic manifold $(\mathcal{M}, \omega, \beta)$

The definition of an observable $(n-1)$-form F, of the bracket and of the observable functionals $\int_\Sigma F$ can be adapted *mutatis mutandis* to the case of a premultisymplectic manifold $(\mathcal{M}, \omega, \beta)$. The difference is that in such a space the dynamical condition $d\mathcal{H}(\xi) = 0$ is empty (think that $\mathcal{M}$ is the level set of some Hamiltonian function $\mathcal{H}$ on a multisymplectic manifold, then the fact that ξ is tangent to this level set forces it to be in the kernel of $d\mathcal{H}$). Hence *any observable $(n-1)$-form is a dynamical one*.

Moreover if ω is exact, i.e. $\omega = d\theta$, we know that Hamiltonian n-curve are critical points of the action (6.11). We can thus see that dynamical observable $(n-1)$-forms correspond to symmetries of the variational problems and the conservation law $dF|_\Gamma = 0$ for a Hamiltonian n-curve Γ is nothing but Noether's first theorem [30, 28]. Indeed for any observable $(n-1)$-form F,

$$L_{\xi_F}\theta = d(\xi_F \lrcorner \theta) + \xi_F \lrcorner d\theta = d(\xi_F \lrcorner \theta) + \xi_F \lrcorner \omega = d(\xi_F \lrcorner \theta - F).$$

Hence $L_{\xi_F}\theta$ is exact, so that ξ_F is a symmetry of the action $\int_\Gamma \theta$ up to a divergence term. The conserved current is just $F|_\Gamma$.

6.1.7 Hamilton–Jacobi equations

The Hamilton–Jacobi equation for a Hamiltonian function $\mathcal{H}$ on a multisymplectic manifold of the form $\Lambda^n T^* \mathcal{Z}$ (or on a submanifold of it) is the following equation on an $(n-1)$-form S on $\mathcal{Z}$ (i.e. a section of the vector bundle $\Lambda^{n-1} T^* \mathcal{Z} \longrightarrow \mathcal{Z}$):

$$\mathcal{H}(z, dS_z) = 0. \tag{6.15}$$

Alternatively the unknown may be chosen to be $\lambda := dS$: we then require that λ is a closed n-form on $\mathcal{Z}$ (or a section λ of $\Lambda^n T^* \mathcal{Z} \longrightarrow \mathcal{Z}$ s.t. $\lambda^* \omega = 0$) which is a solution of $\lambda^* \mathcal{H} = 0$.

Then it for instance $\mathcal{Z} = \mathcal{X} \times \mathcal{Y}$, and if we denote by π the projection from $\mathcal{X} \times \mathcal{Y}$ to $\mathcal{X}$, $\lambda := \pi^* dS$ provides us with a *null Lagrangian functional* $\int_{\mathcal{X}} \lambda$ on $\mathcal{X}$ (i.e. a Lagrangian density which satisfies the Euler–Lagrange equation for any map). In contrast with non relativistic quantum mechanics, the usefulness of this equation in the quantization of fields is not clear for the moment. One of the interests of the Hamilton–Jacobi equation is that it allows one in principle to prove under some circumstances that some solutions of the Euler–Lagrange system of equations are global minimizers, by following a classical strategy designed by K. Weierstrass and D. Hilbert (see [47, 6, 36]). This strategy is the exact analogue in the general theory of calculus of variations of the theory of *calibrations* used in minimal surfaces.

Note that one could impose extra conditions such as requiring that $\lambda = ds^1 \wedge \cdots \wedge ds^n$, where $s^1, \ldots, s^n$ are functions on $\mathcal{Z}$ plus the fact that the graph of λ is foliated by solutions to the Hamilton equations (this provides then a generalization of the picture built by Hamilton in order to reconcile the Fermat principle with the Huygens principle): this was achieved by Carathéodory [2] in his theory (see §6.1.8).

6.1.8 Some historical remarks

The generalization of the Hamilton equations to variational problems with several variables developed first along two directions. One of these is the question of deciding whether a given solution to a variational problem is a minimum of the action functional. This question was answered locally for 1-dimensional variational problems by C.G.J. Jacobi (by following a remark of Legendre) in 1837 [22] by founding a method to check that the second variation is nonnegative which is based on solutions to the so-called Jacobi equation. Note that this method was extended to several variables by A. Clebsch [4] in 1859. Later on a global, nonlinear version of these ideas was developed by K. Weierstrass and D. Hilbert to prove the minimality of some solutions. This theory is connected with another famous work of Jacobi of the same year (1837), who obtained the Hamilton–Jacobi equation [21] by generalizing the work of Hamilton relating the Fermat principle to the Huygens principle. In 1890 V. Volterra wrote two papers [45, 46] where, to my knowledge for the first time[3], two different generalizations of the Hamilton system of equations to variational problems with several variables were proposed. In [46] Volterra extended the Weierstrass–Hilbert theory to variational problems with several variables. This theory was further developed by G. Prange in 1915 [32] and by C. Carathéodory in his book in 1929 [2] and is called today the *Carathéodory* theory. In 1934 H. Weyl [47], inspired by Carathéodory's theory, proposed a variant of it which is based on the same theory as the one proposed by Volterra in [45] and that we described

---

[3] This was followed by a work by L. Koenigsberger [27] in 1901, quoted by T. De Donder in [8], which unfortunately I have difficulties to understand.

in §6.1.1. Today this theory is called the *De Donder–Weyl* one by many authors[4]. Its geometrical framework is the affine multisymplectic manifold $\Lambda_2^n T^* \mathcal{Z}$.

A second direction was the notion of *invariant integrals* due to H. Poincaré [33] and further developed by E. Cartan [3] in 1922: here one emphasizes the relationship of Hamilton equations with the search for invariants which may be functions or differential forms. This point of view is strongly related to the covariant phase space theory (see §6.2.1 below). This theory was developed in full generality by T. De Donder [8] in 1935 and his main contribution was to deduce the extension of the affine ('De Donder–Weyl') theory to Lagrangian densities depending on an arbitrary number of derivatives, i.e. to the theory expounded in §6.1.2. Hence although Weyl's and De Donder's contributions are almost simultaneous they are independent in their inspiration: Weyl's starting point was the so-called Carathéodory theory, motivated by the search for generalizations to several variables of the Hamilton–Jacobi equation, whereas De Donder's starting point was the theory of integral invariants.

The fact that a continuum of different theories may exist for a given variational problem was first understood by T. Lepage [29] in 1936 and completely described by P. Dedecker in 1953 [6]. Today we can picture these various theories as submanifolds of the universal multisymplectic manifold $\Lambda^n T^* \mathcal{Z}$ introduced by J. Kijowski [24] in 1974.

Recently the so-called De Donder–Weyl theory (but that we should call the 'first Volterra theory') has beed studied by many authors starting with the important work by the Polish school around 1970, i.e. by W. Tulczjew, J. Kijowski, W. Szczyrba and later on in many papers which are referred to in e.g. [17, 11]. However the Lepage–Dedecker theory has received much less attention (to my knowledge it was only considered by J. Kijowski [24], F. Hélein, J. Kouneiher [19, 20, 17] and M. Forger, S. V. Romero [10]), probably because of its complexity. The latter theory leads however to interesting phenomena, particularly for gauge theories [19, 17], since first class Dirac constraints simply disappear there.

The modern formulation using the multisymplectic $(n + 1)$-form as the key of the structure of the theory seems to start with the papers of J. Kijowski [23], H. Goldschmidt and P. Sternberg [16] in 1973 and the introduction of observable $(n - 1)$-forms apparentely goes back to the work of K. Gawédski [15] in 1972.

---

[4] Including, in previous papers, the author of this note, who until recently was unaware of the work of Volterra.

6.1.9 An example

Let $\mathcal{X}$ be the n-dimensional Minkowski space-time with coordinates

$$x = (x^0, x^1, \ldots, x^{n-1})$$

and consider the linear Klein–Gordon equation on $\mathcal{X}$:

$$\Box\varphi + m^2\varphi := \frac{\partial^2\varphi}{\partial t^2} - \Delta\varphi + m^2\varphi = 0, \qquad (6.16)$$

where $t = x^0$ and $\Delta := \sum_{i=1}^{n-1} \frac{\partial^2}{(\partial x^i)^2}$. We use the notations $\mathbf{x} := (x^1, \ldots, x^{n-1}) \in \mathbb{R}^{n-1}$ and $x = (x^0, \mathbf{x}) \in \mathbb{R}^n$ and we define the Euclidean scalar product $\mathbf{x} \cdot \mathbf{y} := x^1 y^1 + \cdots + x^{n-1} y^{n-1}$ on $\mathbb{R}^{n-1}$ and the Minkowski product

$$x \cdot y = \eta_{\mu\nu} x^\mu y^\nu = x^0 y^0 - x^1 y^1 - \cdots - x^{n-1} y^{n-1} = x^0 y^0 - \mathbf{x} \cdot \mathbf{y},$$

on $\mathcal{X}$. The multisymplectic formulation of (6.16) takes place in $\mathcal{M} := \Lambda^n T^*(\mathcal{X} \times \mathbb{R})$, equipped with the multisymplectic form

$$\omega := de \wedge \beta + dp^\mu \wedge d\varphi \wedge \beta_\mu.$$

Note that $\omega = d\theta_\lambda$, where

$$\theta_\lambda := e\beta + \lambda p^\mu d\varphi \wedge \beta_\mu - (1 - \lambda)\varphi dp^\mu \wedge \beta_\mu,$$

where $\lambda \in \mathbb{R}$ is a parameter to be fixed later. The Hamiltonian function on $\mathcal{M}$ corresponding to solutions of (6.16) is

$$\mathcal{H}(x, \varphi, e, p) := e + \frac{1}{2}\eta_{\mu\nu} p^\mu p^\nu + \frac{1}{2}m^2\varphi^2.$$

To a solution φ of (6.16) we associate a Hamiltonian n-curve

$$\Gamma = \{(x, \varphi(x), e(x), p(x)) \mid x \in \mathcal{X}\}$$

which satisfies

$$\begin{cases} p^\mu(x) = \eta^{\mu\nu} \dfrac{\partial\varphi}{\partial x^\nu}(x) \\[2mm] e(x) = -\dfrac{1}{2}\eta^{\mu\nu} \dfrac{\partial\varphi}{\partial x^\mu}(x)\dfrac{\partial\varphi}{\partial x^\nu}(x) - \dfrac{1}{2}m^2\varphi(x)^2. \end{cases} \qquad (6.17)$$

We define $\mathcal{E}$ to be the set of Hamiltonian n-curves s.t. for all time t, $x \longmapsto \varphi(t, \mathbf{x})$ is rapidly decreasing at infinity.

We denote by $\mathfrak{P}_{\mathcal{H}}^{n-1}\mathcal{M}$ the set of dynamical observable $(n-1)$-forms F and $\mathfrak{sp}_{\mathcal{H}}\mathcal{M} := \{\xi \mid L_\xi\omega = 0, d\mathcal{H}(\xi) = 0\}$. Note that $(n-1)$-forms F in $\mathfrak{P}_{\mathcal{H}}^{n-1}\mathcal{M}$ are found by looking at vector fields ξ in $\mathfrak{sp}_{\mathcal{H}}\mathcal{M}$ and by solving

$\xi \,\lrcorner\, \omega + dF = 0$. They are of the form

$$F = \zeta \,\lrcorner\, \theta + F_\Phi,$$

where ζ is a vector field on the Minkowski space-time $\mathcal{X}$ which is a generator of the action of the Poincaré group and

$$F_\Phi := \left(p^\mu \Phi(x) - \varphi \eta^{\mu\nu} \frac{\partial \Phi}{\partial x^\nu}(x) \right) \beta_\mu,$$

where Φ is a solution of (6.16). Note that moreover

$$\xi_\Phi := \xi_{F_\Phi} = \Phi(x) \frac{\partial}{\partial \varphi} + \eta^{\mu\nu} \frac{\partial \Phi}{\partial x^\nu}(x) \frac{\partial}{\partial p^\mu} - \left(m^2 \varphi \Phi(x) + p^\mu \frac{\partial \Phi}{\partial x^\mu}(x) \right) \frac{\partial}{\partial e}.$$

In the following we shall denote by

$$P_\mu^{(\lambda)} := \frac{\partial}{\partial x^\mu} \,\lrcorner\, \theta_\lambda$$

and we observe that since $L_{\frac{\partial}{\partial x^\mu}} \theta_\lambda = 0$, we have $dP_\mu^{(\lambda)} + \frac{\partial}{\partial x^\mu} \,\lrcorner\, \omega = 0$. Hence $\xi_{P_\mu^{(\lambda)}} = \frac{\partial}{\partial x^\mu}$.

The brackets of two dynamical observable forms F, $G \in \mathfrak{P}_{\mathcal{H}}^{n-1}\mathcal{M}$ are given as follows: for any pair Φ, Ψ of solutions of (6.16),

$$\{F_\Phi, F_\Psi\} = \eta^{\mu\nu} \left(\frac{\partial \Phi}{\partial x^\nu}(x) \Psi(x) - \Phi(x) \frac{\partial \Psi}{\partial x^\nu}(x) \right) \beta_\mu. \qquad (6.18)$$

We observe that $d\{F_\Phi, F_\Psi\} = 0$. Hence $\left(\mathfrak{P}_{\mathcal{H}}^{n-1}\mathcal{M}, \{\cdot, \cdot\} \right)$ can be understood as a kind of central extension of the Lie algebra $\left(\mathfrak{sp}_{\mathcal{H}}\mathcal{M}, [\cdot, \cdot] \right)$ and the Lie sub algebra spanned by forms F_Φ as an infinite dimensional analogue of the Heisenberg algebra with central charges given by (6.18). Lastly

$$\left\{ P_\mu^{(\lambda)}, F_\Phi \right\} = L_{\frac{\partial}{\partial x^\mu}} F_\Phi - d\left(\frac{\partial}{\partial x^\mu} \,\lrcorner\, F_\Phi \right) = F_{\frac{\partial \Phi}{\partial x^\mu}} - d\left(\frac{\partial}{\partial x^\mu} \,\lrcorner\, F_\Phi \right) \quad (6.19)$$

and $\left\{ P_\mu^{(\lambda)}, P_\nu^{(\lambda)} \right\} = 0$.

For the purpose of quantization we look at functionals of the form $\mathcal{F} = \int_\Sigma F_\Phi$ which are simultaneously eigenvectors of the linear operators

$$\mathcal{F} \longmapsto \left\{ \int_\Sigma P_\mu^{(\lambda)}, \mathcal{F} \right\},$$

for $\mu = 0, \ldots, n-1$. We find by using relation (6.19) that the eigenvector equation reduces to $\frac{\partial \Phi}{\partial x^\lambda} = c\Phi$. This implies (by using the eigenvalue equation for $\mu = 0, \ldots, n-1$) that $\Phi(x) = \alpha e^{ik \cdot x}$. But because Φ should also be a solution of (6.16) we must then have

$$\eta_{\mu\nu} k^\mu k^\nu = (k^0)^2 - |k|^2 = m^2. \qquad (6.20)$$

(We remark that the maps $\{\int_\Sigma P_\lambda^{(\lambda)}, \cdot\}$ play the role of the generators of a Cartan subalgebra.) Let us denote by $\mathcal{C}$ the *mass shell*, i.e. the set of all $k = (k^0, \boldsymbol{k}) \in \mathbb{R}^4$ which are solutions of (6.20). This set actually splits into two connected components according to the sign of k^0: we let $\mathcal{C}^+ := \{k \in \mathcal{C} \mid k^0 > 0\}$. For any $k \in \mathcal{C}^+$ we define

$$\alpha_k := F_{ie^{ik\cdot x}/\sqrt{2\pi}^3} = \frac{i}{\sqrt{2\pi}^3} e^{ik\cdot x} (p^\mu - i\varphi k^\mu) \beta_\mu$$

$$\alpha_k^* := F_{-ie^{-ik\cdot x}/\sqrt{2\pi}^3} = \frac{-i}{\sqrt{2\pi}^3} e^{-ik\cdot x} (p^\mu + i\varphi k^\mu) \beta_\mu.$$

The vector fields associated to these observable forms are:

$$\xi_k := \xi_{\alpha_k} = \frac{e^{ik\cdot x}}{\sqrt{2\pi}^3} \left[i\frac{\partial}{\partial\varphi} - k^\mu \frac{\partial}{\partial p^\mu} + \left(\eta_{\mu\nu} p^\mu k^\nu - im^2\varphi\right) \frac{\partial}{\partial e} \right],$$

$$\xi_k^* := \xi_{\alpha_k^*} = \frac{e^{-ik\cdot x}}{\sqrt{2\pi}^3} \left[-i\frac{\partial}{\partial\varphi} - k^\mu \frac{\partial}{\partial p^\mu} + \left(\eta_{\mu\nu} p^\mu k^\nu + im^2\varphi\right) \frac{\partial}{\partial e} \right].$$

We then define the observable functionals

$$a_k := \int_\Sigma \alpha_k \quad \text{and} \quad a_k^* := \int_\Sigma \alpha_k^*.$$

As the notations suggest these functionals are the classical analogues of respectively the annihilation and the creation operators. The advantage however is that our functionals a_k and a_k^* are independent of the coordinate system. We can choose Σ to be the hyperplane $x^0 = t = 0$ and, for any function f, denote by $f|_0$ the restriction of f to Σ. Then, for any $\Gamma \in \mathcal{E}$ we have

$$a_k(\Gamma) = \frac{i}{\sqrt{2\pi}^3} \int_{\mathbb{R}^3} \left(\frac{\partial\varphi}{\partial t}|_0(\boldsymbol{x}) - ik^0 \varphi|_0(\boldsymbol{x}) \right) e^{-i\boldsymbol{k}\cdot\boldsymbol{x}} d\boldsymbol{x} = i\widehat{\frac{\partial\varphi}{\partial t}|_0}(\boldsymbol{k}) + k^0 \widehat{\varphi|_0}(\boldsymbol{k}),$$

where, for all function ψ on $\mathbb{R}^3$,

$$\widehat{\psi}(\boldsymbol{k}) := \frac{1}{\sqrt{2\pi}^3} \int_{\mathbb{R}^3} \psi(\boldsymbol{x}) e^{-i\boldsymbol{k}\cdot\boldsymbol{x}} d\boldsymbol{x}.$$

Similarly we have:

$$a_k^*(\Gamma) = \frac{-i}{\sqrt{2\pi}^3} \int_{\mathbb{R}^3} \left(\frac{\partial\varphi}{\partial t}|_0(\boldsymbol{x}) + ik^0 \varphi|_0(\boldsymbol{x}) \right) e^{i\boldsymbol{k}\cdot\boldsymbol{x}} d\boldsymbol{x}$$

$$= -i\widehat{\frac{\partial\varphi}{\partial t}|_0}(-\boldsymbol{k}) + k^0 \widehat{\varphi|_0}(-\boldsymbol{k}).$$

Hence we deduce that

$$\widehat{\varphi|_0}(k) = \frac{1}{2k^0}\left(a_k(\Gamma) + a_{\overline{k}}^*(\Gamma)\right) \quad\text{and}\quad \widehat{\frac{\partial\varphi}{\partial t}|_0}(k) = \frac{1}{2i}\left(a_k(\Gamma) - a_{\overline{k}}^*(\Gamma)\right),$$

where $\overline{k} = \overline{(k^0, \boldsymbol{k})} := (k^0, -\boldsymbol{k})$. Thus denoting $d\mu(k) = \frac{1}{2k^0}dk^1dk^2dk^3 = \frac{1}{2k^0}d\boldsymbol{k}$, we have

$$\varphi(0, \boldsymbol{x}) = \frac{1}{\sqrt{2\pi}^3}\int_{\mathbb{R}^3}\frac{1}{2k^0}d\boldsymbol{k}\,e^{i\boldsymbol{k}\cdot\boldsymbol{x}}\left(a_k(\Gamma) + a_{\overline{k}}^*(\Gamma)\right)$$

$$= \frac{1}{\sqrt{2\pi}^3}\int_{\mathcal{C}^+}d\mu(k)\left(a_k(\Gamma)e^{-ik\cdot x} + a_k^*(\Gamma)e^{ik\cdot x}\right)$$

and

$$\frac{\partial\varphi}{\partial t}(0, \boldsymbol{x}) = \frac{-i}{\sqrt{2\pi}^3}\int_{\mathbb{R}^3}\frac{1}{2}d\boldsymbol{k}\,e^{i\boldsymbol{k}\cdot\boldsymbol{x}}\left(a_k(\Gamma) - a_{\overline{k}}^*(\Gamma)\right)$$

$$= \frac{-i}{\sqrt{2\pi}^3}\int_{\mathcal{C}^+}d\mu(k)k^0\left(a_k(\Gamma)e^{-ik\cdot x} - a_k^*(\Gamma)e^{ik\cdot x}\right).$$

Recall that these integrals can be interpreted as integrals over $\mathcal{C}^+$ through the parametrization $\mathbb{R}^3 \ni \boldsymbol{k} \longmapsto (k^0, \boldsymbol{k}) \in \mathcal{C}^+$ and that $d\mu$ is a measure on $\mathcal{C}^+$ invariant by the action of the Lorentz group. Note also that in order to agree with some textbooks one should add an extra factor $\sqrt{k^0}$ inside the integrals. By using the relation (6.18) we obtain, $\forall k, \ell \in \mathcal{C}^+$,

$$\{\alpha_k, \alpha_\ell\} = \frac{-ie^{i(k+\ell)\cdot x}}{(2\pi)^3}(k^\mu - \ell^\mu)\beta_\mu,$$

$$\{\alpha_k^*, \alpha_\ell^*\} = \frac{ie^{-i(k+\ell)\cdot x}}{(2\pi)^3}(k^\mu - \ell^\mu)\beta_\mu,$$

$$\{\alpha_k, \alpha_\ell^*\} = \frac{ie^{i(k-\ell)\cdot x}}{(2\pi)^3}(k^\mu + \ell^\mu)\beta_\mu.$$

These brackets cannot be integrated over the slice $\Sigma := \{x^0 = 0\}$ in the measure theoretical sense[5], but one can make sense of their integrals as distributions over the variables $\boldsymbol{k} \pm \boldsymbol{\ell} \in \mathbb{R}^3$:

$$\{a_k, a_\ell\} = \{a_k^*, a_\ell^*\} = 0, \quad \forall k, \ell \in \mathcal{C}^+,$$

and

$$\{a_k, a_\ell^*\} = i2k^0\delta(\boldsymbol{\ell} - \boldsymbol{k}).$$

---

[5] This contrasts with the integrals $\int_{\Sigma\cap\Gamma}\alpha_k$ and $\int_{\Sigma\cap\Gamma}\alpha_k^*$ which exist if the restrictions to Σ of φ and of its time derivative are Lebesgue integrable.

A way to regularize these operators and their brackets is, by using functions $f, g \in L^2(\mathcal{C}^+)$, to define

$$a_f := \int_{\mathcal{C}^+} d\mu(k) f(k) a_k, \quad \text{and} \quad a_g^* := \int_{\mathcal{C}^+} d\mu(k) g(k) a_k^*.$$

Then

$$\{a_f, a_g^*\} = i \int_{\mathcal{C}^+} d\mu(k) f(k) g(k).$$

6.2 The covariant phase space

6.2.1 A short historical review

The simplest version of the covariant phase space is the set $\mathcal{E}$ of solutions of a Hamiltonian time evolution problem. In this case the Cauchy problem consists in choosing some point M_0 in the ordinary phase space (classically positions and momenta) and some time t_0 and in looking for solutions of the Hamiltonian vector flow which coincide with M_0 at time t_0. This problem has a unique solution in all regular cases and this means that $\mathcal{E}$ is in one to one correspondence with the set of initial data $\{M_0\}$. In other words to each time t_0 there corresponds a natural 'Cauchy coordinates system' on $\mathcal{E}$, which is just the set of initial conditions $\{M_0\}$. The key property is that the Hamiltonian flow preserves the symplectic structure: this means all the symplectic structures induced by these 'Cauchy coordinates systems' on $\mathcal{E}$ coincides. Hence this defines a canonical symplectic structure on $\mathcal{E}$. The substitution of the ordinary phase space by the space of solutions is a classical analogue of the transition from the *Schrödinger picture* to the *Heisenberg picture* in quantum Mechanics: in the Schrödinger picture the dynamics of a particle is described by the evolution of some time dependent 'state' which is represented by a complex line in some complex Hilbert space (the quantum analogue of a point in the ordinary phase space), whereas in the Heisenberg picture the state (still a complex line in a complex Hilbert space) does not evolve with time so that it may be interpreted as a quantum analogue of a solution of the dynamical equations, i.e. of a point in $\mathcal{E}$ (actually more precisely on a Lagrangian submanifold in the phase space, according to A. Weinstein).

In Mechanics this concept is relatively old: the idea of contemplating the space of solutions of a mechanical problem itself probably has roots in the method of the *variation de la constante* of J.L. Lagrange and the notion of 'Lagrange bracket' is very close to the symplectic structure on the phase space. The observation that this space carries an intrinsic symplectic structure was clearly formulated by H. Poincaré [33] in his theory of *invariant integrals* (*invariants intégraux*) and later further developed by E. Cartan [3] and fully recognized by J.M. Souriau [39]. T. De Donder [7] extended the notion of integral invariant to variational problems with several variables, being hence very to from the notion of covariant phase space in this context, although it is not clear that he realized it. Actually it seems difficult to decide when the concept of covariant phase space in field theory emerged. My own guess is that such an idea could also have been inspired by quantum field theory, since it may be thought as the classical analogue of the Heisenberg picture in quantum fields theory. First known works in this direction are the R.E. Peierls bracket in 1952 [31], followed by the paper of I. Segal [37] in 1960. Peierls defined a bracket on the set of solutions to a relativistic hyperbolic wave equation which can be understood now as the restriction of the Poisson bracket associated to the covariant phase structure on a certain class of functionals on the phase space. Segal proved that the set of solutions of a non linear field relativistic wave equation precisely carries a symplectic structure and proposed to use this symplectic structure to quantize fields (and his paper is among the ones at the origin of the geometric quantization method). This idea was later developed in a more and more general framework by P. L. García [12] in 1968, García and A. Pérez-Rendón [13] in 1971, H. Goldschmidt, S. Sternberg [16] in 1973. In my opinion, the more accomplished presentation is the one by J. Kijowski and W. Szczyrba [25] in 1976, which gives the first elementary but general presentation of this structure, by using the multisymplectic formalism.

A more recent appearance of this idea can be found in the papers by C. Crnkovic and E. Witten [5] and by G. Zuckerman [48], where the authors were apparently unaware of the previous references and have rediscovered this principle, being guided by the concept of the variational bicomplex of F. Takens [40] and the work of A.M. Vinogradov [41]. This was followed by several developments in the physical (e.g. [9]) and the mathematical literature, where this principle is often referred to as the Witten covariant phase space. A general presentation in the framework of the *secondary calculus* of Vinogradov was done by E. Reyes [34] and L. Vitagliano in [42] and in relation to multisymplectic geometry (as in the present paper) by M. Forger and S.V. Romero in [10].

6.2.2 The basic principle

We expound here briefly the principle of the covariant phase space using the multsymplectic formalism. Our presentation will be heuristic and we refer to [25, 18] for details. We assume that we are given a premultisymplectic manifold

$(\mathcal{M}, \omega, \beta)$ (see §6.1.4) and, as in §6.1.5, that ω is exact, i.e. $\omega = d\theta$, for some n-form θ. We note $\mathcal{E}$ the set of Hamiltonian n-curves in $(\mathcal{M}, \omega, \beta)$, i.e. the set of oriented n-dimensional submanifolds $\Gamma \subset \mathcal{M}$ which satisfy (6.10). Given some $\Gamma \in \mathcal{E}$, the tangent space[6] to $\mathcal{E}$ at Γ represents the set of infinitesimal deformations $\delta\Gamma$ of Γ which preserves the equation (6.10). Such a deformation $\delta\Gamma$ can be represented by a vector field ξ tangent to $\mathcal{M}$ defined along Γ, i.e. a section over Γ of $j_\Gamma^* T\mathcal{M}$, which is the pull-back image of the tangent bundle $T\mathcal{M}$ by the embedding map $j_\Gamma : \Gamma \longrightarrow \mathcal{M}$. Given $\delta\Gamma$, the vector field ξ is of course not unique, since for any tangent vector field ζ on Γ (i.e. a section of the subbundle $T\Gamma \subset j_\Gamma^* T\mathcal{M}$), $\xi + \zeta$ represents also $\delta\Gamma$. If so we write:

$$\delta\Gamma = \int_\Gamma \xi = \int_\Gamma \xi + \zeta.$$

Moreover the condition on $\delta\Gamma$ of being tangent to $\mathcal{E}$ forces ξ to be a solution of the *Jacobi equation*:

$$\forall v \in \mathcal{C}^\infty(\mathcal{M}, T_\mathcal{M}\mathcal{M}), \quad \left(v \lrcorner L_\xi \omega\right)|_\Gamma = 0. \tag{6.21}$$

Note that, although ξ is not a vector field defined on $\mathcal{M}$ (neither on a neighbourhood of Γ in $\mathcal{M}$) but only on Γ, one can make sense of $L_\xi \omega|_\Gamma$ because Γ is a solution of (6.10).

Then for any slice Σ (see §6.1.6), any $\Gamma \in \mathcal{E}$ and $\delta\Gamma \in T_\Gamma\mathcal{E}$, we define

$$\Theta_\Gamma^\Sigma(\delta\Gamma) := \int_{\Sigma \cap \Gamma} \xi \lrcorner \theta,$$

where ξ is a section of $j_\Gamma^* T\mathcal{M}$ over Γ s.t. $\delta\Gamma = \int \xi$ and $\xi \lrcorner \theta$ is the interior product of θ by ξ. This hence defines a 1-form Θ^Σ on $\mathcal{E}$

The dependence of Θ^Σ on Σ

This is the first natural question. For that purpose let us consider a smooth 1-parameter family of slices $(\Sigma_t)_t$ and compute the derivative:

$$\frac{d}{dt}\left(\Theta_\Gamma^{\Sigma_t}(\delta\Gamma)\right) = \frac{d}{dt}\left(\int_{\Sigma_t \cap \Gamma} \xi \lrcorner \theta\right) = \int_{\Sigma_t \cap \Gamma} L_{\frac{\partial}{\partial t}}(\xi \lrcorner \theta)$$

$$= \int_{\Sigma_t \cap \Gamma} \frac{\partial}{\partial t} \lrcorner d(\xi \lrcorner \theta) + d\left(\xi \wedge \frac{\partial}{\partial t} \lrcorner \theta\right).$$

But $d(\xi \lrcorner \theta) = L_\xi \theta - \xi \lrcorner d\theta$ and thus

$$\frac{\partial}{\partial t} \lrcorner d(\xi \lrcorner \theta) = \frac{\partial}{\partial t} \lrcorner \left(L_\xi \theta\right) - \frac{\partial}{\partial t} \lrcorner \xi \lrcorner \omega.$$

---

[6] Note that since $\mathcal{E}$ may not be a manifold in general, the usual definition of a tangent space should be replaced by a suitable notion, see [25, 18].

However we can assume w.l.g. (see [18]) that the vector fields $\frac{\partial}{\partial t}$ and ξ admit extensions s.t. $\left[\xi, \frac{\partial}{\partial t}\right] = 0$. Then the preceding relation gives us

$$\frac{\partial}{\partial t} \lrcorner d(\xi \lrcorner \theta) = L_\xi \left(\frac{\partial}{\partial t} \lrcorner \theta\right) - \xi \wedge \frac{\partial}{\partial t} \lrcorner \omega.$$

Hence

$$\frac{d}{dt}\left(\Theta_\Gamma^{\Sigma_t}(\delta\Gamma)\right) = \int_{\Sigma_t \cap \Gamma} L_\xi \left(\frac{\partial}{\partial t} \lrcorner \theta\right) - \int_{\Sigma_t \cap \Gamma} \xi \wedge \frac{\partial}{\partial t} \lrcorner \omega$$
$$+ \int_{\Sigma_t \cap \Gamma} d\left(\xi \wedge \frac{\partial}{\partial t} \lrcorner \theta\right). \qquad (6.22)$$

First let us consider a smooth curve $s \longmapsto \Gamma_s \in \mathcal{E}$ s.t. $\Gamma_0 = \Gamma$ and $\frac{d\Gamma_s}{dt} = \delta\Gamma$. Then the first term in the r.h.s. of (6.22) is equal to

$$\int_{\Sigma_t \cap \Gamma} L_\xi \left(\frac{\partial}{\partial t} \lrcorner \theta\right) = \frac{d}{ds}\left(\int_{\Sigma_t \cap \Gamma_s} \frac{\partial}{\partial t} \lrcorner \theta\right)\bigg|_{s=0} = \delta S_\Gamma^{\frac{d\Sigma_t}{dt}}(\delta\Gamma),$$

where we have posed:

$$S^{\frac{d\Sigma_t}{dt}}(\Gamma) := \int_{\Sigma_t \cap \Gamma} \frac{\partial}{\partial t} \lrcorner \theta.$$

Second we can assume w.l.g. (see [18]) that we can choose $\frac{\partial}{\partial t}$ in such a way that it is tangent to Γ. Let $(X_2, \ldots, X_n)$ be a system of tangent vectors on Γ s.t. $\forall t$, $\forall M \in \Sigma_t \cap \Gamma$, $(X_2(M), \ldots, X_n(M))$ is a basis of $T_M(\Sigma_t \cap \Gamma)$ and $(\frac{\partial}{\partial t}(M), X_2(M), \ldots, X_n(M))$ is a basis of $T_M\Gamma$. Then if ψ is a n-volume form on Γ s.t. $\psi(\frac{\partial}{\partial t}, X_2, \ldots, X_n) = 1$, the second term in the r.h.s. of (6.22) reads

$$- \int_{\Sigma_t \cap \Gamma} \xi \wedge \frac{\partial}{\partial t} \lrcorner \omega = - \int_{\Sigma_t \cap \Gamma} \omega\left(\xi, \frac{\partial}{\partial t}, X_2, \ldots, X_n\right) \psi$$

and vanishes because of the Hamilton equations (6.10). Lastly we assume that *the restriction of ξ to $\Sigma_t \cap \Gamma$ has compact support or is rapidly decreasing*: this occurs for instance if the Hamilton system encodes hyperbolic wave equations, if Σ is a level hypersurface of some time coordinate and if we impose that the Hamiltonian n-curves in $\mathcal{E}$ have a prescribed behaviour at infinity in space for all time. Then the last term in the r.h.s. of (6.22) vanishes. Then Relation (6.22) can be rewritten

$$\frac{d}{dt}\left(\Theta_\Gamma^{\Sigma_t}(\delta\Gamma)\right) = \delta S_\Gamma^{\frac{d\Sigma_t}{dt}}(\delta\Gamma), \quad \forall \delta\Gamma \subset T_\Gamma \mathcal{E}$$

or

$$\frac{d}{dt}\left(\Theta^{\Sigma_t}\right) = \delta S^{\frac{d\Sigma_t}{dt}}. \qquad (6.23)$$

We can also define the functional

$$S_{\Sigma_1}^{\Sigma_2}(\Gamma) := \int_{\Gamma \cap \{t_1 \le t \le t_2\}} \theta,$$

which represents the 'action' between the slices $\Sigma_1 := \{t = t_1\}$ an $\Sigma_2 := \{t = t_2\}$. Then $S_{\Sigma_1}^{\Sigma_2}(\Gamma) = \int_{t_1}^{t_2} S^{\frac{d\Sigma_t}{dt}}(\Gamma)dt$ and thus we deduce by integrating (6.23) over $[t_1, t_2]$ that

$$\Theta^{\Sigma_2} - \Theta^{\Sigma_1} = \delta S_{\Sigma_1}^{\Sigma_2}. \tag{6.24}$$

The symplectic form

In view of the preceding we are led to the conclusion that, although the 1-form Θ^Σ depends on Σ, its differential $\delta\Theta^\Sigma$ does not depend on Σ since (6.24) tells us that $\Theta^{\Sigma_2} - \Theta^{\Sigma_1}$ is an exact form. Of course one should be careful in using the identity $\delta \circ \delta = 0$ since $\mathcal{E}$ is not a smooth manifold (see [18] for a rigorous proof that $\delta\Theta^\Sigma$ does not depend on Σ). All this motivates the definition of the following 2-form on $\mathcal{E}$:

$$\Omega := \delta\Theta^\Sigma.$$

We will prove that Ω has the following expression: $\forall \delta_1\Gamma, \delta_2\Gamma \in T_\Gamma\mathcal{E}$,

$$\Omega_\Gamma(\delta_1\Gamma, \delta_2\Gamma) = \int_{\Sigma \cap \Gamma} \xi_1 \wedge \xi_2 \lrcorner \omega, \tag{6.25}$$

where ξ_1, ξ_2 are sections over Γ of $j_\Gamma^* T\mathcal{M}$ s.t. $\delta_1\Gamma = \int_\Gamma \xi_1$ and $\delta_2\Gamma = \int_\Gamma \xi_2$. To prove (6.25) we need to compute $\delta\Theta_\Gamma^\Sigma(\delta_1\Gamma, \delta_2\Gamma)$. For that purpose we first assume that we can extend the two tangent vectors $\delta_1\Gamma$ and $\delta_2\Gamma$ to commuting vector fields on $\mathcal{E}$ around Γ (actually we can assume that $[\xi_1, \xi_2] = 0$). Then

$$\delta\Theta_\Gamma^\Sigma(\delta_1\Gamma, \delta_2\Gamma) = \delta_1\Gamma \cdot \Theta_\Gamma^\Sigma(\delta_2\Gamma) - \delta_2\Gamma \cdot \Theta_\Gamma^\Sigma(\delta_1\Gamma) - \Theta_\Gamma^\Sigma([\delta_1\Gamma, \delta_2\Gamma])$$

$$= \delta_1\Gamma \cdot \left(\int_{\Sigma \cap \Gamma} \xi_2 \lrcorner \theta \right) - \delta_2\Gamma \cdot \left(\int_{\Sigma \cap \Gamma} \xi_1 \lrcorner \theta \right).$$

Thus

$$\delta\Theta_\Gamma^\Sigma(\delta_1\Gamma, \delta_2\Gamma) = \int_{\Sigma \cap \Gamma} L_{\xi_1} (\xi_2 \lrcorner \theta) - \int_{\Sigma \cap \Gamma} L_{\xi_2} (\xi_1 \lrcorner \theta).$$

We use then the following identity (see [18]): for any pair of vector fields X_1 and X_2 and for any p-form β,

$$L_{X_1}(X_2 \lrcorner \beta) - L_{X_2}(X_1 \lrcorner \beta) = X_1 \wedge X_2 \lrcorner d\beta + [X_1, X_2] \lrcorner \beta + d(X_1 \wedge X_2 \lrcorner \beta).$$

Setting $X_1 = \xi_1, X_2 = \xi_2$ and $\beta = \theta$, we obtain using $[\xi_1, \xi_2] = 0$ and $d\theta = \omega$ that $L_{\xi_1}(\xi_2 \lrcorner \theta) - L_{\xi_2}(\xi_1 \lrcorner \theta) = \xi_1 \wedge \xi_2 \lrcorner \omega - d(\xi_1 \wedge \xi_2 \lrcorner \theta)$. Thus

$$\delta \Theta_\Gamma^\Sigma (\delta_1 \Gamma, \delta_2 \Gamma) = \int_{\Sigma \cap \Gamma} \xi_1 \wedge \xi_2 \lrcorner \omega - d(\xi_1 \wedge \xi_2 \lrcorner \theta). \qquad (6.26)$$

Hence if we assume that the restriction of ξ_1 and ξ_2 to $\Sigma_t \cap \Gamma$ has compact support or is rapidly decreasing (as in the preceding paragraph) we obtain (6.25).

Hence we conclude that, under some hypotheses, one can endow the set $\mathcal{E}$ of solutions to the Hamilton equations with a symplectic form Ω given by (6.25). This form does depend not on Σ under the condition that the boundary terms $\int_{\Sigma_t \cap \Gamma} d\left(\xi \wedge \frac{\partial}{\partial t} \lrcorner \theta \right)$ in (6.22) and $- \int_{\Sigma \cap \Gamma} d(\xi_1 \wedge \xi_2 \lrcorner \theta)$ in (6.26) vanish. This means that, on each slice Σ, the Jacobi vector fields ξ, ξ_1, ξ_2 decrease sufficiently rapidly at infinity. Such a condition is true if, for instance, the manifold $\mathcal{X}$ is a Lorentzian manifold, the slice Σ is (a lift of) a spacelike hypersurface of $\mathcal{X}$ and we impose in the definition of $\mathcal{E}$ that all Hamiltonian n-curves Γ in $\mathcal{E}$ are asymptotic to a given 'ground state' Hamiltonian n-curve Γ_0 at infinity on each slice Σ.

With such a symplectic structure Ω on $\mathcal{E}$ we can define a Poisson bracket on real-valued functionals on $\mathcal{E}$, which is nothing but (6.13).

6.2.3 A geometric view of the proof

We can give an alternative proof of Relation (6.24) with a more geometric flavor. We will be even more heuristic, however the validity of our argument is strongly based on the fact that the Lagrangian action can be represented by (6.11). For that purpose imagine that our problem models a hyperbolic time evolution problem and that there are well-defined notions of time and space coordinates on $\mathcal{M}$ (as is the case for any wave equation on a curved space-time).

Consider a Hamiltonian n-curve Γ and let Γ' be another Hamiltonian n-curve, which we suppose to be close to Γ. More precisely we assume that $\Gamma' = \Gamma + \varepsilon \delta \Gamma + o(\varepsilon)$, where $\varepsilon > 0$ is a small parameter : by this condition we mean that there exists a vector field $\xi \in \mathcal{C}^\infty(\mathcal{M}, T\mathcal{M})$ s.t. $\delta \Gamma = \int_\Gamma \xi$ and Γ' is the image of Γ by the flow map $e^{\varepsilon \xi}$. We also assume that, for all 'time', Γ' is asymptotic to Γ at infinity in space. Let Σ_1 and Σ_2 be two slices, assume that these slices are space-like hypersurfaces and, in order to fix ideas, we suppose that Σ_2 is in the future of Σ_1. These slices cross transversally Γ and Γ' and we denote by σ_1 (resp. σ_2) the piece of Σ_1 (resp. Σ_2) which is enclosed by the intersections with Γ and Γ' (see the picture). We also denote by Γ_- the part of Γ which is in the past of Σ_1, by Γ_+ the part of Γ which is in the future of Σ_2 and

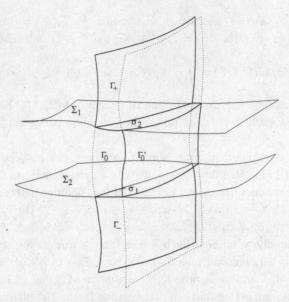

Figure 6.1 A geometric comparison of $\Theta^{\Sigma_1}(\delta\Gamma)$ with $\Theta^{\Sigma_2}(\delta\Gamma)$

by Γ_0' the part of Γ' which is between Σ_1 and Σ_2 (see again the picture). Lastly we consider the (not necessarily Hamiltonian) n-curve Γ_ε, which is the union of Γ_-, σ_1, Γ_0', σ_2 and Γ_+. Of course Γ_ε is not smooth, but it can be approached by a sequence of smooth n-curves, so that the following makes sense. We also endow Γ_ε with the orientation which agrees with that of Γ on $\Gamma_- \cup \Gamma_+$ and with that Γ' on Γ_0'.

Let us use the fact that Γ is a Hamiltonian n-curve, hence a critical point of (6.11). This implies that

$$\int_{\Gamma_\varepsilon} \theta = \int_\Gamma \theta + o(\varepsilon). \qquad (6.27)$$

However the l.h.s. of (6.27) can be decomposed as

$$\int_{\Gamma_\varepsilon} \theta = \int_{\Gamma_-} \theta + \int_{\sigma_1} \theta + \int_{\Gamma_0'} \theta + \int_{\sigma_2} \theta + \int_{\Gamma_+} \theta,$$

whereas its r.h.s. is

$$\int_\Gamma \theta + o(\varepsilon) = \int_{\Gamma_-} \theta + \int_{\Gamma_0} \theta + \int_{\Gamma_+} \theta + o(\varepsilon),$$

where Γ_0 is the part of Γ between Σ_1 and Σ_2. Hence (6.27) reduces to

$$\int_{\sigma_1} \theta + \left(\int_{\Gamma_0'} \theta - \int_{\Gamma_0} \theta \right) + \int_{\sigma_2} \theta = o(\varepsilon). \tag{6.28}$$

We now recognize that, on the one hand, $\int_{\sigma_1} \theta = \varepsilon \Theta^{\Sigma_1}(\delta\Gamma) + o(\varepsilon)$, $\int_{\sigma_2} \theta = -\varepsilon \Theta^{\Sigma_2}(\delta\Gamma) + o(\varepsilon)$ (the sign being due to the orientation of σ_2). On the other hand $\int_{\Gamma_0} \theta = S_{\Sigma_1}^{\Sigma_2}(\Gamma_0)$ and $\int_{\Gamma_0'} \theta = S_{\Sigma_1}^{\Sigma_2}(\Gamma_0') = S_{\Sigma_1}^{\Sigma_2}(\Gamma_0 + \varepsilon\delta\Gamma_0) + o(\varepsilon)$. Hence (6.28) gives us

$$\varepsilon \Theta^{\Sigma_1}(\delta\Gamma) + \varepsilon \left(\delta S_{\Sigma_1}^{\Sigma_2} \right)_\Gamma (\delta\Gamma) - \varepsilon \Theta^{\Sigma_2}(\delta\Gamma) = o(\varepsilon).$$

Thus by dividing by ε and letting ε tend to 0, we recover (6.11).

6.3 Geometric quantization

We address here the question of building a geometric quantization scheme, or at least a prequantization scheme for fields based on the covariant phase space structure. This was more or less the programme envisioned by G. Segal in 1960 [37]. We present here an attempt at that by using the multisymplectic theory on a very elementary example, which is the one presented in §6.1.9.

Canonical vector fields on the set of solutions $\mathcal{E}$

We can associate to each $F \in \mathfrak{P}_{\mathcal{H}}^{n-1}\mathcal{M}$ a tangent vector field Ξ_F on $\mathcal{E}$ wich is given by

$$\forall \Gamma \in \mathcal{E}, \quad \Xi_F(\Gamma) := \int_\Gamma \xi_F.$$

In the case of the Klein–Gordon equation (6.16) it is interesting to represent solutions in $\mathcal{E}$ by local coordinates. The most convenient way is based on the Fourier transform: any Hamiltonian n-curve Γ is characterized by a solution φ to (6.16) and by writing

$$\varphi(x) = \frac{1}{\sqrt{2\pi}^3} \int_{C^+} d\mu(k) \left(u_k e^{-ik \cdot x} + u_k^* e^{ik \cdot x} \right), \tag{6.29}$$

we get formally a map

$$\mathcal{E} \longrightarrow \mathbb{C}^{C^+} \times \mathbb{C}^{C^+}$$
$$\varphi \longmapsto (u_k, u_k^*)_{k \in C^+}.$$

Note that the image of $\mathcal{E}$ is characterized by the reality condition $\overline{u_k} = u_k^*$, $\forall k \in C^+$. We can obviously extend this map to the complexification $\mathcal{E}^{\mathbb{C}}$ of $\mathcal{E}$ and then this map is an isomorphism.

The creation and annihilation canonical transformations

Now given some function $f \in L^2(C^+)$ consider

$$\alpha_f := \int_{C^+} d\mu(k) f(k) \alpha_k = \frac{i}{\sqrt{2\pi}^3} \int_{C^+} d\mu(k) e^{ik \cdot x} f(k)(p^\mu - i\varphi k^\mu) \beta_\mu.$$

(Note that the observable functional a_f defined in §6.1.9 is obtained by integration of α_f over a slice.) Then

$$\xi_f := \xi_{\alpha_f} = \frac{i}{\sqrt{2\pi}^3} \int_{C^+} d\mu(k) e^{ik \cdot x} f(k) \left(ik^\mu \frac{\partial}{\partial p^\mu} - (m^2\varphi + i\eta_{\mu\nu} p^\mu k^\nu) \frac{\partial}{\partial e} + \frac{\partial}{\partial \varphi} \right)$$

is completely characterized by the fact that it preserves ω and $d\mathcal{H}$ and through its action on φ:

$$d\varphi \left(\xi_f \right) = \frac{i}{\sqrt{2\pi}^3} \int_{C^+} d\mu(k) e^{ik \cdot x} f(k).$$

We can easily integrate ξ_f on $\mathcal{M}$ and its action on $\mathcal{F}$:

$$U(s, \varphi) = \varphi + s \frac{i}{\sqrt{2\pi}^3} \int_{C^+} d\mu(k) e^{ik \cdot x} f(k).$$

In terms of the coordinates $(u_k, u_k^*)_{k \in C^+}$ it gives:

$$U(s, u_k, u_k^*) = (u_k, u_k^* + is f(k)).$$

Hence we can symbolically denote

$$\Xi_f := \Xi_{\alpha_f} = i \int_{C^+} f(k) \frac{\partial}{\partial u_k^*}.$$

There is no integration measure used here, the sign $\int$ stands uniquely for summing linearly independent vectors: the meaning is that

$$\Xi_f \left(\frac{1}{\sqrt{2\pi}^3} \int_{C^+} \frac{dk}{2k^0} \left(u_k e^{-ik \cdot x} + u_k^* e^{ik \cdot x} \right) \right) = \frac{1}{\sqrt{2\pi}^3} \int_{C^+} d\mu(k) i f(k) e^{ik \cdot x}.$$

A completely analogous computation can be done for

$$\alpha_g^* := \int_{C^+} d\mu(k) g(k) \alpha_k^* = \frac{-i}{\sqrt{2\pi}^3} \int_{C^+} d\mu(k) e^{-ik \cdot x} g(k)(p^\mu + i\varphi k^\mu) \beta_\mu,$$

where $g \in L^2(\mathcal{C}^+)$. Denoting $\xi_g^* := \xi_{\alpha_g^*}$ we have

$$d\varphi\left(\xi_g^*\right) = \frac{-i}{\sqrt{2\pi}^3} \int_{\mathcal{C}^+} d\mu(k) e^{-ik \cdot x} g(k).$$

Hence $\Xi_g^* := \Xi_{\alpha_g^*}$ is given by

$$\Xi_g^* = -i \int_{\mathcal{C}^+} g(k) \frac{\partial}{\partial u_k}.$$

Spacetime translations

We now look at the canonical vector fields on $\mathcal{E}$ associated with spacetime translations P_ζ, where ζ is constant vector field on $\mathcal{X}$. We recall that $\xi_{P^{(\lambda)}} = \zeta$. We must understand the induced vector field Ξ_ζ on $\mathcal{F}$. Let $U(s, \cdot)$ be the flow mapping of the vector field $\zeta \colon U(s, x, \varphi, e, p) = (s, x + \zeta, \varphi, e, p)$. Then the image of

$$\Gamma = \{(x, \varphi(x), e(x), p(x)) \mid x \in \mathcal{X}\} \subset \mathcal{E}$$

by $U(x, \cdot)$ is

$$\Gamma_s = \{(x, \varphi(x - s\zeta), e_s(x), p_s(x)) \mid x \in \mathcal{X}\},$$

where the value of $e_s(x)$ and $p_s(x)$ is completely determined by the constraint that $\Gamma_s \subset \mathcal{E}$ and by the knowledge of $\varphi(x - s\zeta)$. This can be proved by a simple change of variable. Similarly we determine the action of Ξ_ζ on the coordinates $(u_k, u_k^*)_{k \in \mathcal{C}^+}$ by computing its action on φ:

$$(\Xi_\zeta \varphi)(x) = \frac{d}{ds} \left(\frac{1}{\sqrt{2\pi}^3} \int_{\mathcal{C}^+} d\mu(k) \left(u_k e^{-ik \cdot (x - s\zeta)} + u_k^* e^{ik \cdot (x - s\zeta)} \right) \right) \Bigg|_{s=0}$$

$$= \frac{1}{\sqrt{2\pi}^3} \int_{\mathcal{C}^+} d\mu(k) \left(ik \cdot \zeta u_k e^{-ik \cdot (x - s\zeta)} - ik \cdot \zeta u_k^* e^{ik \cdot (x - s\zeta)} \right)$$

$$= i \left[\left(\int_{\mathcal{C}^+} k \cdot \zeta \left(u_k \frac{\partial}{\partial u_k} - u_k^* \frac{\partial}{\partial u_k^*} \right) \right) \varphi \right](x).$$

Hence

$$\Xi_\zeta = i \int_{\mathcal{C}^+} k \cdot \zeta \left(u_k \frac{\partial}{\partial u_k} - u_k^* \frac{\partial}{\partial u_k^*} \right).$$

Geometric prequantization

We recall very briefly the prequantization scheme due to B. Kostant and J.-M. Souriau (generalizing previous constructions by B.O. Koopman, L. Van Hove and I. Segal, see [26, 38]). We let $(\mathcal{M}, \omega)$ be a simply connected symplectic manifold and we assume for simplicity that there exists a 1-form θ with $\omega = d\theta$. We consider the trivial bundle $\mathcal{L} := \mathcal{M} \times \mathbb{C}$ and denote by $\Gamma(\mathcal{M}, \mathcal{L})$ the set of square integrable sections of $\mathcal{L}$. Using θ we can define a Hermitian connection ∇ acting on $\Gamma(\mathcal{M}, \mathcal{L})$ by

$$\forall \xi \in \Gamma(\mathcal{M}, T\mathcal{M}), \forall \psi \in \Gamma(\mathcal{M}, \mathcal{L}), \quad \nabla_\xi \psi = \xi \cdot \psi - \frac{i}{\hbar}\theta(\xi)\psi.$$

Then to each function $F \in \mathcal{C}^\infty(\mathcal{M}, \mathbb{R})$ we associate the operator $\widehat{F}$ acting on $\Gamma(\mathcal{M}, \mathcal{L})$

$$\widehat{F}\psi = F\psi + \frac{\hbar}{i}\nabla_{\xi_F}\psi = (F - \theta(\xi_F))\,\psi + \frac{\hbar}{i}\xi_F \cdot \psi,$$

where $dF + \xi_F \lrcorner \omega = 0$. This construction is called the prequantization of $(\mathcal{M}, \omega)$. For instance if $(\mathcal{M}, \omega) = (\mathbb{R}^{2n}, dp_i \wedge dq^i)$, then $\omega = d\theta$, with $\theta = p_i dq^i$ and $\widehat{q^i} = q^i + i\hbar\frac{\partial}{\partial p_i}$ and $\widehat{p_i} = -i\hbar\frac{\partial}{\partial q^i}$. Of course one needs further restrictions in order to recover an irreducible representation of the Heisenberg algebra (and hence the standard quantization): this will be the purpose of introducing a polarization and a tensorization of the line bundle $\mathcal{L}$ with the bundle of half volume forms transversal to the leaves of the polarization (see [26, 38]).

We will propose an extension of this procedure to our setting, concerned with the quantization of fields. We consider the trivial bundle $\mathcal{L} := \mathcal{E}^{\mathbb{C}} \times \mathbb{C}$ over $\mathcal{E}^{\mathbb{C}}$, where $\mathcal{E}^{\mathbb{C}}$ is the complexification of the set of solutions to the Klein–Gordon equation (6.16) as before. On the set $\Gamma(\mathcal{E}^{\mathbb{C}}, \mathcal{L})$ of smooth sections of $\mathcal{L}$ (we are here relatively vague about the meaning of "smooth") we define a notion of covariant derivative along any vector field of the type Ξ_F, where $F \in \Gamma(\mathcal{E}^{\mathbb{C}}, \mathcal{L})$ by

$$\forall \psi \in \Gamma(\mathcal{E}^{\mathbb{C}}, \mathcal{L}), \quad \nabla_{\Xi_F}\psi := \Xi_F \cdot \psi - \frac{i}{\hbar}\left(\int_\Sigma \xi_F \lrcorner \theta\right)\psi,$$

where $\theta = \theta_\lambda$. Then we define the prequantization of $F \in \mathfrak{P}_{\mathcal{H}}^{n-1}\mathcal{M}$ to be the operator acting on $\Gamma(\mathcal{E}^{\mathbb{C}}, \mathcal{L})$ by:

$$\widehat{F}\psi := \left(\int_\Sigma F\right)\psi + \frac{\hbar}{i}\nabla_{\Xi_F}\psi = \frac{\hbar}{i}\Xi_F \cdot \psi + \left(\int_\Sigma F - \xi_F \lrcorner \theta\right)\psi.$$

Prequantization of the creation and annihilation observables
We look here for the expressions of the prequantization of a_f and a_g^* given in §6.1.9. We first note the fact that if φ is given in terms of $(u_k, u_k^*)_{k \in \mathcal{C}^+}$ by (6.29),

then

$$\frac{1}{\sqrt{2\pi}^3} \int_{\mathbb{R}^n} e^{-ik \cdot x} \varphi(0, \boldsymbol{x}) d\boldsymbol{x} = \frac{u_k + u_{\overline{k}}^*}{2k^0},$$

and

$$\frac{1}{\sqrt{2\pi}^3} \int_{\mathbb{R}^n} e^{-ik \cdot x} p^0(0, \boldsymbol{x}) d\boldsymbol{x} = \frac{u_k - u_{\overline{k}}^*}{2i},$$

where $\overline{k} = (k^0, -\boldsymbol{k})$. We deduce the following

$$\int_{\Sigma \cap \Gamma} \alpha_f = \frac{1}{\sqrt{2\pi}^3} \int_{\mathbb{R}^n} d\boldsymbol{x} \int_{C^+} \frac{d\boldsymbol{k}}{2k^0} e^{-ik \cdot x} f(k) \left(\varphi(0, \boldsymbol{x}) k^0 + i p^0(0, \boldsymbol{x}) \right)$$

$$= \int_{C^+} \frac{d\boldsymbol{k}}{2k^0} f(k) \left(\frac{u_k + u_{\overline{k}}^*}{2} + \frac{u_k - u_{\overline{k}}^*}{2} \right)$$

$$= \int_{C^+} \frac{d\boldsymbol{k}}{2k^0} f(k) u_k.$$

Similarly

$$\int_{\Sigma \cap \Gamma} \alpha_g^* = \int_{C^+} \frac{d\boldsymbol{k}}{2k^0} g(k) u_k^*.$$

We moreover observe that

$$\xi_f \lrcorner \theta = \frac{1}{\sqrt{2\pi}^3} \int_{C^+} \frac{d\boldsymbol{k}}{2k^0} e^{ik \cdot x} \frac{f(k)}{2} (\varphi k^\mu + i p^\mu) \beta_\mu = \frac{\alpha_f}{2},$$

and similarly $\xi_g^* \lrcorner \theta = \frac{\alpha_g^*}{2}$. Hence

$$\int_{\Sigma \cap \Gamma} \xi_f \lrcorner \theta = \int_{C^+} \frac{d\boldsymbol{k}}{2k^0} \frac{f(k)}{2} u_k \quad \text{and} \quad \int_{\Sigma \cap \Gamma} \xi_g^* \lrcorner \theta = \int_{C^+} \frac{d\boldsymbol{k}}{2k^0} \frac{g(k)}{2} u_k^*.$$

Using the previous results we can now express, for $\psi \in \Gamma(\mathcal{E}^{\mathbb{C}}, \mathcal{L})$,

$$\nabla_{\Xi_f} \psi := \Xi_f \cdot \psi - \frac{i}{\hbar} \left(\int_\Sigma \xi_f \lrcorner \theta \right) \psi$$

$$= i \int_{C^+} f(k) \frac{\partial \psi}{\partial u_k^*} - \frac{i}{\hbar} \left(\int_{C^+} \frac{d\boldsymbol{k}}{2k^0} \frac{f(k)}{2} u_k \right) \psi$$

and

$$
\begin{aligned}
\nabla_{\Xi_g^*}\psi &:= \Xi_g^* \cdot \psi - \frac{i}{\hbar}\left(\int_\Sigma \xi_g^* \lrcorner\, \theta\right)\psi \\
&= -i\int_{C^+} g(k)\frac{\partial\psi}{\partial u_k} - \frac{i}{\hbar}\left(\int_{C^+}\frac{d\boldsymbol{k}}{2k^0}\frac{g(k)}{2}u_k^*\right)\psi.
\end{aligned}
$$

For the prequantizations we obtain:

$$
\widehat{a}_f\psi = \hbar\int_{C^+} f(k)\frac{\partial\psi}{\partial u_k^*} + \left(\int_{C^+}\frac{d\boldsymbol{k}}{2k^0}\frac{f(k)}{2}u_k\right)\psi,
$$

and

$$
\widehat{a}_g^*\psi = -\hbar\int g(k)\frac{\partial\psi}{\partial u_k} + \left(\int_{C^+}\frac{d\boldsymbol{k}}{2k^0}\frac{g(k)}{2}u_k^*\right)\psi.
$$

We observe that we have formally $[\widehat{a}_f, \widehat{a}_{f'}] = [\widehat{a}_g^*, \widehat{a}_{g'}^*] = 0$ and

$$
[\widehat{a}_f, \widehat{a}_g^*] = \hbar\int_{C^+}\frac{d\boldsymbol{k}}{2k^0}f(k)g(k).
$$

Prequantization of the stress-energy tensor

It relies on finding the prequantization of $P_\zeta^{(\lambda)} = \zeta^\mu P_\mu^{(\lambda)}$. In principle one should compute the functionals of $\int_\Sigma P_\zeta^{(\lambda)}$ and $\int_\Sigma \xi_{P_\zeta^{(\lambda)}} \lrcorner\, \theta_\lambda$. But as observed in the previous section we have $P_\zeta^{(\lambda)} = \zeta \lrcorner\, \theta_\lambda = \xi_{P_\zeta^{(\lambda)}} \lrcorner\, \theta_\lambda$ because $Lie_\zeta \theta_\lambda = 0$. Hence $\int_\Sigma P_\zeta^{(\lambda)} - \xi_{P_\zeta^{(\lambda)}} \lrcorner\, \theta_\lambda = 0$ and so the prequantization of $P_\zeta^{(\lambda)}$ is just

$$
\widehat{P^{(\lambda)}}_\zeta\psi = \frac{\hbar}{i}\Xi_\zeta \cdot \psi = \hbar\int_{C^+} k\cdot\zeta\left(u_k\frac{\partial}{\partial u_k} - u_k^*\frac{\partial}{\partial u_k^*}\right)\psi.
$$

Note that if we need to compute $\int_\Sigma P_\zeta^{(\lambda)}$, it is more suitable to set $\lambda = 1$, since it gives then the standard expression for the stress-energy tensor. For instance, if $\zeta = \frac{\partial}{\partial x^0}$

$$
-\int_{\Sigma\cap\Gamma} P_0^{(1)} = \int_{\mathbb{R}^{n-1}} d\boldsymbol{x}\left(\frac{(p^0)^2}{2} + \sum_{i=1}^3\frac{(p^i)^2}{2} + m^2\frac{\varphi^2}{2}\right)
$$

gives the total energy in the frame associated with the coordinates x^μ.

Introducing a polarization

We choose to impose the extra condition $\nabla_{\Xi_g^*}\psi = 0$, $\forall g$ (covariantly antiholo-morphic sections), which gives us:

$$\psi(u_k, u_k^*) = h(u_k^*)\exp\left(-\frac{1}{2\hbar}\int_{C^+}\frac{d\boldsymbol{k}}{2k^0}u_k u_k^*\right) = h(u_k^*)|0\rangle.$$

The advantage of this choice is that all observable functionals (creation and annihilation, energy and momentum) are at most linear in the variables (u_k, u_k^*), so we do not need to use the Blattner–Kostant–Sternberg correction for these operators [26, 38]. As a result $\widehat{P}|0\rangle = 0$, so that the energy of the vacuum vanishes without requiring normal ordering. However we did not take into account the metaplectic correction, which requires a slight change of the connection: were we to do so we would find that the vacuum has infinite energy (as in the standard quantization scheme), which can be removed by a normal ordering procedure. The mysterious thing here (as was already observed) is that by ignoring the metaplectic correction (which however is fundamental for many reasons) we do not need the normal ordering correction.

References

[1] J. C. Baez, C. L. Rogers, *Categorified Symplectic Geometry and the String Lie 2-algebra*, Homology, Homotopy Appl. 12 (2010), no. 1, 221–236. arXiv:0901. 4721.

[2] C. Carathéodory, *Variationsrechnung und partielle Differentialgleichungen erster Ordnung*, Teubner, Leipzig (reprinted by Chelsea, New York, 1982); Acta litt. ac scient. univers. Hungaricae, Szeged, Sect. Math., 4 (1929), p. 193.

[3] E. Cartan, *Leçons sur les invariants intégraux*, Hermann, 1922.

[4] A. Clebsch, *Ueber die zweite Variation vielfache Integralen*, J. reine angew. Math. 56 (1859), 122–148.

[5] C. Crnkovic, E. Witten, *Covariant description of canonical formalism in geometrical theories*, in *Three hundred years of gravitation* (S. W. Hawking and W. Israel, eds.), Cambridge University Press, Cambridge, 1987, 676–684; E. Witten, *Interacting field theory of open supertrings*, Nucl. Phys. B 276 (1986), 291–324.

[6] P. Dedecker, *Calcul des variations, formes différentielles et champs géodésiques*, in *Géométrie différentielle*, Colloq. Intern. du CNRS LII, Strasbourg 1953, Publ. du CNRS, Paris, 1953, p. 17–34; *On the generalization of symplectic geometry to multiple integrals in the calculus of variations*, in *Differential Geometrical Methods in Mathematical Physics*, eds. K. Bleuler and A. Reetz, Lect. Notes Maths. vol. 570, Springer-Verlag, Berlin, 1977, p. 395–456.

[7] T. De Donder, *Introduction à la théorie des invariants intégraux*, Bull. Acad. Roy. Belgique (1913), 1043–1073.

[8] T. De Donder, *Théorie invariantive du calcul des variations*, Gauthiers-Villars, Paris, 1930.

[9] B.P. Dolan, K.P. Haugh, *A co-variant approach to Ashtekar's canonical gravity*, Class. Quant. Gravity, Vol. 14, N. 2, (1997), 477–488 (12).

[10] M. Forger, S. V. Romero, *Covariant Poisson bracket in geometric field theory*, Commun. Math. Phys. 256 (2005), 375–410.

[11] M. Forger, L. Gomes, *Multisymplectic and polysymplectic structures on fiber bundles*, preprint arXiv:0708.1586.

[12] P. L. García, *Geometría simplética en la teoria de campos*, Collect. Math. 19, 1–2, 73, 1968.

[13] P. L. García, A. Pérez-Rendón, *Symplectic approach to the theory of quantized fields, I*, Commun. Math. Phys. 13 (1969), 24–44 and —, *II*, Arch. Rational Mech. Anal. 43 (1971), 101–124.

[14] M.J. Gotay, J. Isenberg, J.E. Marsden (with the collaboraton of R. Montgomery, J. Śnyatycki, P.B. Yasskin), *Momentum maps and classical relativistic fields, Part I/covariant field theory*, preprint arXiv/physics/9801019

[15] K. Gawędski, *On the generalization of the canonical formalism in the classical field theory*, Rep. Math. Phys. No 4, Vol. 3 (1972), 307–326.

[16] H. Goldschmidt, S. Sternberg, *The Hamilton–Cartan formalism in the calculus of variations*, Ann. Inst. Fourier Grenoble 23, 1 (1973), 203–267.

[17] F. Hélein, *Hamiltonian formalisms for multidimensional calculus of variations and perturbation theory*, in *Noncompact problems at the intersection of geometry, analysis, and topology*, Contemp. Math., 350 (2004), 127–147.

[18] F. Hélein, *The use of the covariant phase space on non nonlinear fields*, in preparation.

[19] F. Hélein, J. Kouneiher, *Covariant Hamiltonian formalism for the calculus of variations with several variables: Lepage–Dedecker versus De Donder–Weyl*, Adv. Theor. Math. Phys. 8 (2004), 565–601.

[20] F. Hélein, J. Kouneiher, *The notion of observable in the covariant Hamiltonian formalism for the calculus of variations with several variables*, Adv. Theor. Math. Phys. 8 (2004), 735–777.

[21] C.G.J. Jacobi, *Ueber die Reduction der Integration des partiellen Differentialgleichungen erster Ornung zwischen irgend einer Zahl Variabeln auf die Integration eines einzigen Systemes gewöhnlicher Differentialgleichungen*, J. reine angew. Math. 17 (1837), 68–82.

[22] C.G.J. Jacobi, *Zur Theorie des Variations-Rechnung und des Differential-Gleichungen*, J. reine angew. Math. 17 (1837), 97–162.

[23] J. Kijowski, *A finite dimensional canonical formalism in the classical field theory*, Commun. Math. Phys. 30 (1973), 99–128.

[24] J. Kijowski, *Multiphase spaces and gauge in the calculus of variations*, Bull. de l'Acad. Polon. des Sci., Série sci. Math., Astr. et Phys. XXII (1974), 1219–1225.

[25] J. Kijowski, W. Szczyrba, *A canonical structure for classical field theories*, Commun. Math Phys. 46 (1976), 183–206.

[26] A.A. Kirillov, *Geometric quantization*, in *Dynamical systems IV*, V.I. Arnol'd, S.P. Novikov, eds., Springer-Verlag, 1990.

[27] L. Koenigsberger, *Die Prinzipien der Mechanik für mehrere Variable*, Sitzungsberichte Akad. Wiss. Berlin, Bd. XLVI, 14 nov. 1901, 1108; *Die Prinzipien der*

Mechanik für mehrere unavhängige Variable, J. Reine Angw. Math., Bd. 124 (1902), 202–277.

[28] Y. Kosmann-Schwarzbach, *Les théorèmes de Noether – Invariance et lois de conservation au XXe siècle*, Les éditions de l'Ecole Polytechnique, 2004.

[29] T. Lepage, *Sur les champs géodésiques du calcul des variations*, Bull. Acad. Roy. Belg., Cl. Sci. 27 (1936), 716–729, 1036–1046.

[30] E. Noether, *Invariante Variationsprobleme*, Nachrichten von der Königlichen Gesellschaft des Wissenschaften su Göttingen, Mathematisch-physikalische Kalsse, 1918, p. 235–257.

[31] R.E. Peierls, *The commutation laws of relativistic field theory*, Proc. Roy. Soc. London, Ser. A, Vol. 214, No. 1117 (1952), 143–157.

[32] G. Prange, *Die Hamilton–Jacobische Theorie für Doppelintegrale*, Diss. Göttingen, 1915.

[33] H. Poincaré, *Les méthodes nouvelles de la mécanique céleste*, t. III, Paris, Gauthier-Villars, 1899.

[34] E.G. Reyes, *On covariant phase space and the variational bicomplex*, Int. J. Theor. Phys., Vol. 43, No. 5 (2004), 1267–1286.

[35] C. Rovelli, *A note on the foundation of relativistic mechanics – II: Covariant Hamiltonian general relativity*, in Topics in Mathematical Physics, General Relativity and Cosmology, H Garcia-Compean, B Mielnik, M Montesinos, M Przanowski editors, pg 397, (World Scientific, Singapore 2006). arXiv:gr-qc/0202079

[36] H. Rund, *The Hamilton–Jacobi theory in the calculus of variations, its role in mathematics and physics*, Krieger Pub. 1973.

[37] I. Segal, *Quantization of nonlinear systems*, J. Math. Phys. vol. 1, N. 6 (1960), 468–488.

[38] J. Śnyatycki, *Geometric quantization and quantum Mechanics*, Appl. Math. Sci. 30, Springer-Verlag 1980.

[39] J.-M. Souriau, *Structure des systèmes dynamiques*, Dunod, Paris, 1970.

[40] F. Takens, *A global formulation of the inverse problem of the calculus of variations*, J. Diff. Geom. 14 (1979), 543–562.

[41] A.M. Vinogradov, *The C-spectral sequence, Lagrangian formalism, and conservations laws, I and II*, J. Math. Anal. Appl. 100 (1984), 1–40 and 41–129.

[42] L. Vitagliano, *Secondary calculus and the covariant phase space*, J. Geom. Phys. 59 (2009), no. 4, 426–447.

[43] L. Vitagliano, *The Lagrangian–Hamiltonian Formalism for Higher Order Field Theories*, J. Geom. Phys. 60 (2010), no. 6-8, 857–873. arXiv:0905.4580.

[44] L. Vitagliano, *Partial Differential Hamiltonian Systems*, preprint arXiv:0903.4528

[45] V. Volterra, *Sulle equazioni differenziali che provengono da questiono di calcolo delle variazioni*, Rend. Cont. Acad. Lincei, ser. IV, vol. VI, 1890, 42–54.

[46] V. Volterra, *Sopra una estensione della teoria Jacobi–Hamilton del calcolo delle varizioni*, Rend. Cont. Acad. Lincei, ser. IV, vol. VI, 1890, 127–138.

[47] H. Weyl, *Geodesic fields in the calculus of variation for multiple integrals*, Ann. Math. (3) 36 (1935), 607–629.

[48] G. Zuckerman, *Action functional and global geometry*, in *Mathematical aspects of string theory*, S.T. Yau, eds., Advanced Series in Mathematical Physics, vol 1, World Scientific, 1987, 259–284.

Author's address:

Institut de Mathématiques de Jussieu,
UMR CNRS 7586 Université Denis Diderot Paris 7,
175 rue du Chevaleret,
75013 Paris, France
helein@math.jussieu.fr

7

Nonnegative curvature on disk bundles

LORENZ J. SCHWACHHÖFER

7.1 Introduction

The search for manifolds of nonnegative curvature[1] is one of the classical problems in Riemannian geometry. While general obstructions are scarce, there are relatively few general classes of examples and construction methods. Hence, it is unclear how large one should expect the class of closed manifolds admitting a nonnegatively curved metric to be. For a survey of known examples, see e.g. [12].

Apart from taking products, there are only two general methods to construct new nonnegatively curved metrics out of given spaces. One is the use of *Riemannian submersions* which do not decrease curvature by O'Neill's formula. The other is the glueing of two manifolds (which we call *halves*) along their common boundary. Typically, the boundary of each half is assumed to be totally geodesic or, slightly more restrictively, a *collar metric*. This in turn implies by the *Soul theorem* ([2]) that each half is the total space of a disk bundle over a totally geodesic closed submanifold. In addition, the glueing map of the two boundaries must be an isometry.

While many examples can be constructed by such a glueing, its application is still limited. On the one hand, there is not too much known on the question of which disk bundles over a nonnegatively curved compact manifold admit collar metrics of nonnegative curvature, and on the other hand, even if such metrics exist, the metric on the boundary is not arbitrary. Thus, glueing together two such disk bundles to a nonnegatively curved closed manifold is possible in special situations only.

For instance, if the disk bundle is *homogeneous*, then there always exist invariant nonnegatively curved collar metrics. However, the metric on the

---

[1] Throughout this article, the term "curvature" refers to the sectional curvature.

boundary of such a collar metric is restricted due to the existence of certain parallel Killing fields by a result of Perelman ([7]).

In this article, we will give a survey of known examples and describe some recent results which illustrate the difficulty in finding metrics on disk bundles which are suitable for this glueing construction.

7.2 Normal homogeneous metrics and Cheeger deformations

Let G be a compact Lie group with Lie algebra $\mathfrak{g}$, and let g_0 be a biinvariant metric on G, i.e., a metric for which all left and right translations are isometries. It is well known that each compact Lie group carries such a metric, and its curvature is given by

$$sec_{g_0}(x, y) = \frac{1}{4} \|[x, y]\|_{g_0}^2 \geq 0$$

for each orthonormal pair $x, y \in T_g G = dL_g \mathfrak{g} \cong \mathfrak{g}$. Let $H \subset G$ be a closed subgroup, and consider the compact homogeneous space $M := G/H$. Then there is a unique G-invariant metric on G/H which by abuse of notation we also denote by g_0, for which the canonical projection $G \to G/H$ becomes a Riemannian submersion. This metric on G/H is called a *normal homogeneous metric*. Evidently, a normal homogeneous metric always has nonnegative curvature by O'Neill's formula.

Let (M, g) be a Riemannian manifold, and suppose that G acts isometrically on M. The action map

$$G \times M \longrightarrow M$$

may be regarded as a principal G-bundle with the free action of G on $G \times M$ given by

$$k \star (g, p) := (gk^{-1}, k \cdot p)$$

for $g, k \in G$ and $p \in M$. Evidently, this action is isometric with respect to the product of a biinvariant metric on G and the given metric on M.

Definition 7.2.1 Let (M, g) be a Riemannian manifold, and let G be a Lie group with a biinvariant metric g_0 which acts isometrically on M. Then for $\lambda > 0$, the metric g_λ on M for which the action map

$$(G \times M, \lambda g_0 \oplus g) \longrightarrow (M, g_\lambda)$$

becomes a Riemannian submersion is called a *Cheeger deformation of g*.

Since Cheeger deformations are Riemannian submersions and hence curvature non-decreasing by O'Neill's formula, it follows that g_λ has nonnegative curvature whenever g does. Moreover, g_λ may have nonnegative curvature even if g has some negative curvature.

Proposition 7.2.2 *Let g be a G-invariant metric on M and let g_λ denote the Cheeger deformation of g for $\lambda > 0$. Then*

(i) $\lim_{\lambda \to \infty} g_\lambda = g$.

(ii) $(g_\lambda)_\mu = g_{\frac{\lambda\mu}{\lambda+\mu}}$. *That is, the Cheeger deformation of a Cheeger deformation is again a Cheeger deformation.*

(iii) *If g_{λ_0} has nonnegative curvature for some $\lambda_0 > 0$, then so does g_λ for any $\lambda < \lambda_0$.*

(iv) *If $M = G/H$ is a homogeneous space with a G-invariant metric g, then $\lim_{\lambda \to 0} \frac{1}{\lambda} g_\lambda = g_0$, where g_0 is the normal homogeneous metric on G/H induced by the biinvariant metric g_0 on G.*

Proof Let $p \in M$, and let $N := G \cdot p \subset M$ be the G-orbit through p. We decompose the tangent space g-orthogonally as

$$T_p M = T_p N \oplus S_p.$$

We write $N = G/H$ with $H := Stab(p) \subset G$, and choose the orthogonal decomposition of the Lie algebra $\mathfrak{g} = \mathfrak{h} \oplus \mathfrak{m}$, so that $\mathfrak{m} \cong T_p N$ with the identification $x \mapsto (x^*)_p$, where x^* denotes the action field of $x \in \mathfrak{g}$. Let $\varphi : \mathfrak{m} \to \mathfrak{m}$ be the self-adjoint map for which

$$g(x^*, y^*)_p = g_0(\varphi(x), y),$$

where g_0 is the given biinvariant inner product on $\mathfrak{g}$. The tangent space of the fiber of the submersion $G \times M \to M$ at (e, p) is given by $\{(-x, (x^*)_p) \mid x \in \mathfrak{g}\}$, hence its orthogonal complement with respect to $(\lambda g_0) \oplus g$ is

$$\mathcal{H}_{(e,p)} = \{(\varphi(x), \lambda(x^*)_p) \mid x \in \mathfrak{m}\} \oplus S_p.$$

Therefore, the horizontal lift of $(x^*)_p$ is $(\varphi(\hat{x}), \lambda(\hat{x}^*)_p)$, where $\hat{x} := (\lambda + \varphi)^{-1}(x)$. It follows that for all $\lambda > 0$ we have

$$g_\lambda|_{S_p} = g|_{S_p}, \quad g_\lambda(S_p, T_p N) = 0 \quad \text{and} \quad g_\lambda(x^*, y^*) = g_0(\lambda\varphi(\lambda + \varphi)^{-1}(x), y)$$

for all $x, y \in \mathfrak{m}$. From this, the first and the fourth assertion on the limits of g_λ as $\lambda \to \infty$ and $\lambda \to 0$ follow. The second statement follows since

$$\mu(\lambda\varphi(\lambda + \varphi)^{-1})(\mu + (\lambda\varphi(\lambda + \varphi)^{-1}))^{-1} = \frac{\mu\lambda}{\mu+\lambda}\varphi\left(\frac{\mu\lambda}{\mu+\lambda} + \varphi\right)^{-1},$$

and the third follows since for $\lambda < \lambda_0$ we can write $\lambda = \frac{\mu\lambda_0}{\mu+\lambda_0}$ for $\mu := \frac{\lambda\lambda_0}{\lambda_0-\lambda} > 0$ and then apply the preceding statement and the fact that a Cheeger deformation of a nonnegatively curved metric is again nonnegatively curved. $\square$

7.3 Homogeneous metrics of nonnegative curvature

By virtue of Proposition 7.2.2, we should consider Cheeger deformations with small parameters $\lambda > 0$ when searching for nonnegatively curved metrics. Moreover, since the G-orbits of M are homogeneous spaces, the last statement of Proposition 7.2.2 implies that the metrics on the orbits approach a biinvariant metric for Cheeger deformations with small $\lambda > 0$. Therefore, it is important to investigate the question of which G-invariant metrics on a homogeneous space close to the normal homogeneous one have nonnegative curvature.

In general, this question is too hard to answer. But we can give a partial answer if G/H is the total space of a homogeneous fibration.

Theorem 7.3.1 *([9]) Let $H \subset K \subset G$ be compact Lie groups with Lie algebras $\mathfrak{h} \subset \mathfrak{k} \subset \mathfrak{g}$, and consider the homogeneous fibration $K/H \hookrightarrow G/H \to G/K$. Moreover, denote the orthogonal decompositions with respect to some biinvariant metric g_0 on $\mathfrak{g}$ as*

$$\mathfrak{g} = \mathfrak{h} \oplus \mathfrak{p} = \mathfrak{k} \oplus \mathfrak{s} \ and \ \mathfrak{k} = \mathfrak{h} \oplus \mathfrak{m} \tag{7.1}$$

(i) *The metric on G/H obtained from the normal homogeneous one by shrinking the fibers of this fibration is always nonnegatively curved.*

(ii) *The metric on G/H obtained from the normal homogeneous one by enlarging the fibers of this fibration by a factor $(1 + \varepsilon)$ is nonnegatively curved for $\varepsilon > 0$ sufficiently small if and only if there is a constant $C > 0$ such that for all $X, Y \in \mathfrak{p}$,*

$$|[X^m, Y^m]^m| \le C \cdot |[X, Y]|, \tag{7.2}$$

where the superscripts refer to the decomposition (7.1).

In particular, if the fiber K/H is a symmetric space, then $[X^m, Y^m]^m = 0$, so that in this case, the fiber can always be enlarged while maintaining nonnegative curvature. All metrics on G/H in this theorem are induced by the left-invariant metric given by the inner product

$$g_t := t g_0|_{\mathfrak{k}} + g_0|_{\mathfrak{s}} \tag{7.3}$$

on $\mathfrak{g}$; indeed, shrinking (respectively, enlarging) the fibers corresponds to the case $t < 1$ (respectively, $t > 1$).

Proof For the first statement, we observe that applying the Cheeger deformation to the action of K on G by right multiplication, we obtain that g_λ is the left-invariant metric which on $\mathfrak{g} = T_e G$ is given as

$$g_\lambda = \frac{\lambda}{1 + \lambda} g_0|_{\mathfrak{k}} + g_0|_{\mathfrak{s}},$$

hence this metric on G is nonnegatively curved, and so is the induced metric on G/H, which corresponds to scaling the fibers by $t := \lambda/(1 + \lambda) \in (0, 1)$.

The second part follows from a more careful investigation of the curvature formula for invariant metrics on homogeneous spaces which we shall not present here. The details may be found in [9]. $\qquad\square$

If H is the trivial group so that $G/H = G$, then (7.2) in the above theorem is equivalent to a simpler criterion. Namely, we have the following

Theorem 7.3.2 *([8]) The left-invariant metric on G induced by (7.3) has nonnegative curvature for some $t > 1$ if and only if the semi-simple part of $\mathfrak{k}$ is an ideal of $\mathfrak{g}$. In this case, the metric has nonnegative curvature for all $t \leq 4/3$.*

7.4 Collar metrics of nonnegative curvature

Let M be a manifold with boundary $N := \partial M$. A Riemannian metric g on M is called a *collar metric*, if there is a neighborhood of N which is isometric to $([0, \varepsilon) \times N, dt^2 + g_N)$ for some $\varepsilon > 0$ and some metric g_N on N. Evidently, if M_1, M_2 are manifolds with $\partial M_1 = \partial M_2 =: N$, then collar metrics g_i on M_i whose restriction to the boundary N are isometric can be glued together by an isometry to give a smooth metric on

$$M := M_1 \cup_N M_2.$$

This simple principle has proven very useful to construct metrics of nonnegative curvature, as the following examples due to J. Cheeger illustrate.

Theorem 7.4.1 *([1]) Let $\hat{M}^n$ be a compact rank-one symmetric space (CROSS), and let $M := \hat{M}^n \setminus B_\varepsilon(p)$ be the complement of a small open ball. Then there is a nonnegatively curved collar metric on M such that the metric on $\partial M = S^{n-1}$ is the round metric.*

Corollary 7.4.2 *([1]) Let M_1^n and M_2^n be CROSSes of equal dimension. Then both $M_1 \# M_2$ and $M_1 \# \overline{M}_2$ admit nonnegatively curved metrics.*

The corollary is an immediate consequence of the theorem, since the glueing in this case corresponds to taking the connected sum, and changing the orientation on one of the summands before glueing is possible. In order to prove the theorem, Cheeger used the fact that M is always a homogeneous disk bundle over a CROSS of lower dimension, and he was able to perform the construction of the metric in a fairly explicit way.

If M is a manifold with boundary $N = \partial M$ which admits a nonnegatively curved collar metric, then the *Soul Theorem* ([2]) implies that there is a totally geodesic submanifold $S \subset M$, called the *soul of M*, such that M is the total space of a unit disk bundle $D \to S$, whence $N \to S$ is a sphere bundle.

Given a disk bundle $M \to S$, it is thus a reasonable question to ask if M admits a nonnegatively curved collar metric and if so, what are the possible restrictions of such a metric to $N = \partial M$. If $M \to S$ carries such a metric, then evidently it can be extended to a complete nonnegatively curved metric on the corresponding vector bundle $\hat{M} \to S$ by defining the metric to be the product $([r_0, \infty) \times N, dt^2 + g_N)$ on $\hat{M} \backslash M$. The converse of this statement also holds:

Theorem 7.4.3 *([6]) Let $(\hat{M}, g)$ be a complete open manifold of nonnegative curvature, so that by the soul theorem $\hat{M}$ is the total space of a vector bundle $\hat{M} \to S$. Then there is a nonnegatively curved collar metric on the unit disk bundle $M \subset \hat{M}$.*

As an immediate consequence of this result, we may conclude the following.

Corollary 7.4.4 *Let $\hat{M}_k$, $k = 1, 2$ be open manifolds admitting a complete nonnegatively curved metric, and let $M_k \subset \hat{M}_k$ be the corresponding unit disk bundles. Then there is a nonnegatively curved metric on the (closed) manifold*

$$\partial(M_1 \times M_2).$$

Proof We may write $\partial(M_1 \times M_2) = (M_1 \times \partial M_2) \cup_{\partial M_1 \times \partial M_2} (\partial M_1 \times M_2)$. Now consider the nonnegatively curved collar metrics g_k on M_k whose existence is guaranteed by Theorem 7.4.3. Then the products of these metrics induce nonnegatively curved collar metrics on $M_1 \times \partial M_2$ and $\partial M_1 \times M_2$ whose boundary is isometric to $(\partial M_1 \times \partial M_2, g_1 + g_2)$, so that these metrics may be glued together.   □

7.5 Bundles with normal homogeneous collar

A special class of vector bundles which we wish to consider are *homogeneous vector bundles and disk bundles*. These are bundles of the form

$$\hat{M} := G \times_K V \text{ and } M := G \times_K \overline{B_R(0)} \subset G \times_K V, \text{ respectively,}$$

where $K \subset G$ acts on the Euclidean vector space V by some orthogonal representation $K \to O(V)$ which is transitive on the unit sphere $S^V \subset V$, so that $S^V = K/H$ for some subgroup $H \subset K$.

Note that $\hat{M}$ is the quotient of $G \times V$ by the free left action of K given by $k \star (g, v) := (gk^{-1}, k \cdot v)$. Thus, any K-invariant nonnegatively curved metric on $G \times V$ induces a nonnegatively curved metric on $\hat{M}$, and this metric on $M \subset \hat{M}$ is a collar metric if the metric on $G \times \overline{B_R(0)} \subset G \times V$ is.

If follows that each homogeneous disk bundle has G-invariant nonnegatively curved collar metrics: choose on $G \times \overline{B_R(0)}$ the direct sum of a nonnegatively curved left invariant, Ad_K-invariant metric on G and a nonnegatively curved K-invariant collar metric on $\overline{B_R(0)}$, such as a "*cigar metric*", i.e., a nonnegatively curved $O(V)$-invariant metric which is a cylinder outside a compact set. Clearly, such metrics exist. However, the induced metric on the boundary G/H is in general *not* normal homogeneous, and this limits the possibilities of glueing two such bundles.

Definition 7.5.1 Let $M := G \times_K \overline{B_R(0)} \subset G \times_K V \longrightarrow G/K$ be a homogeneous disk bundle with $K \to O(V)$ acting transitively on the sphere $S^V = K/H$. A metric on M is called a *normal homogeneous collar metric* if a neighborhood of the boundary $\partial M \cong G/H$ is G-equivariantly isometric to $([0, \varepsilon) \times G/H, dt^2 + g_0)$, where g_0 is a normal homogeneous metric on G/H.

Let us suppose that the metric on M is a submersion of $(G \times \overline{B_R(0)}, g_\mathfrak{g} + g_V)$ where $g_\mathfrak{g}$ is a left-invariant Ad_K-invariant metric on G and g_V is a K-invariant collar metric on $\overline{B_R(0)}$ whose boundary metric on $K/H = S^V \subset V$ is induced by a left-invariant metric $g_\mathfrak{k}$ on K. Then the metric on the boundary $\partial M \cong G/H$ is induced by the Riemannian submersion

$$(G \times K, g_\mathfrak{g} + g_\mathfrak{k}) \longrightarrow G \longrightarrow G/H. \qquad (7.4)$$

It is now of interest to see how these metrics on G and $\overline{B_R(0)} \subset V$ can be chosen such that the submersion (7.4) induces a normal homogeneous metric on G/H and hence a normal homogeneous collar metric on M. For the metric on $\overline{B_R(0)} \subset V$, we may use the following result.

Theorem 7.5.2 *([11]) Let $K \to O(V)$ be a representation of a Lie group K on a Euclidean vector space V which is transitive on the unit sphere $S^V \subset V$. Then there is a nonnegatively curved collar metric on $\overline{B_R(0)} \subset V$ which is $Norm_{O(V)}K$-invariant and is normal homogeneous on $S^V = K/H$.*

Let $L \to O(V)$ be a representation which is transitive on the unit sphere $S^V \subset V$, and extend this to a representation $K \times L \to O(V)$, i.e., K maps to the centralizer $Z_{O(V)}(L)$. Since the normal homomogeneous collar metric on V from the Theorem 7.5.2 is $(K \times L)$-invariant, there is an induced nonnegatively curved normal homogeneous collar metric on

$$(G \times L) \times_{K \times L} V \longrightarrow G/K,$$

which as a bundle can be regarded as $G \times_K V \to G/K$, where K acts on V as a subgroup of $Z_{O(V)}(L) \subset O(V)$. Such a homogeneous bundle is called *essentially trivial*. Thus, all essentially trivial homogeneous disk bundles carry a nonnegatively curved collar metric whose boundary is normal homogeneous with respect to the action of $G \times L$.

The following gives an exhaustive description of all essentially trivial bundles which are not trivial.

Example 7.5.3

(i) $\mathbb{Z}_2$-**quotients**. Let $K \lhd K' \subset G$ be such that $K'/K \cong \mathbb{Z}_2$, and let $L \to O(V)$ be a representation which is transitive on the unit sphere. Moreover, define the representation $K' \to K'/K \cong \{\pm Id\} \subset O(V)$. Then the essentially trivial bundle

$$(G \times L) \times_{K' \times L} V \cong G \times_{K'} V \longrightarrow G/K'$$

is the quotient of the trivial bundle $G/K \times V$ by $\mathbb{Z}_2 \cong K'/K$ which acts on G/K from the right and on V as $\{\pm Id\}$.

(ii) **Rank-one bundles**. If $\dim V = 1$, i.e., $V = \mathbb{R}$, then $O(V) \cong \mathbb{Z}_2$, hence any homogeneous bundle of rank one is of the form $G \times_{K'} V \to G/K'$ with a surjective map $K' \to O(V) \cong \mathbb{Z}_2$. The kernel of this map is a normal subgroup $K \lhd K'$ with $K'/K \cong \mathbb{Z}_2$. Thus, this is a special case of the preceding, i.e., we can write any non-trivial rank one homogeneous bundle in the form

$$(G \times \mathbb{Z}_2) \times_{K' \times \mathbb{Z}_2} V \cong G \times_{K'} V \to G/K',$$

which is the quotient $(G/K \times V)/\mathbb{Z}_2$ as above.

(iii) S^1-**quotients**. Similarly, if V is a complex hermitean vector space and $L \to U(V)$ is transitive on the unit sphere, then we consider subgroups $K \lhd K' \subset G$ such that $K'/K = S^1$. Then the essentially trivial bundle

$$(G \times L) \times_{K' \times L} V \cong G \times_{K'} V \longrightarrow G/K'$$

is the quotient of the trivial bundle $G/K \times V$ by the action of $S^1 = K'/K$ on G/K from the right and on V by scalar multiplication by $S^1 \subset \mathbb{C}$.

(iv) $Sp(1)$-**quotients**. Likewise, let V be a quaternionic hermitean vector space and suppose that $L \to Sp(V)$ is surjective and hence transitive on the unit sphere. Again, consider subgroups $K \lhd K' \subset G$ such that $K'/K = Sp(1)$. Then the essentially trivial bundle

$$(G \times L) \times_{K' \times L} V \cong G \times_{K'} V \longrightarrow G/K'$$

is the quotient of the trivial bundle $G/K \times V$ by the action of $Sp(1) = K'/K$ on G/K from the right and on V by scalar multiplication by $Sp(1) \subset \mathbb{H}$.

Thus, all these bundles admit nonnegatively curved normal homogeneous collar metrics. Note that this description of essentially trivial bundles is exhaustive, since by the classification of transitive actions of a Lie group L on a sphere $S^V \subset V$ the centralizer $Z_{O(V)}(L) \subset O(V)$ is either trivial or one of $\mathbb{Z}_2$, S^1 or $Sp(1)$.

By virtue of Theorem 7.5.2, we may assume that $g_{\mathfrak{k}}$ in (7.4) is a biinvariant metric. If we choose $g_{\mathfrak{g}}$ as a biinvariant metric such that $g_{\mathfrak{k}} = (g_{\mathfrak{g}})|_{\mathfrak{k}}$, then it follows that the induced metric on G/H is obtained by shrinking the fiber of the normal homogeneous metric of the fibration $S^V = K/H \hookrightarrow G/H \to G/K$. In particular, this metric on the boundary G/H is *not* normal homogeneous.

In order to achieve a normal homogeneous boundary metric on G/H, we have to define the metric on $\mathfrak{g} = \mathfrak{k} \oplus \mathfrak{s}$ by

$$g_{\mathfrak{g}} := (1 + \varepsilon)(g_0)|_{\mathfrak{k}} + (g_0)|_{\mathfrak{s}}, \qquad (7.5)$$

where g_0 is a biinvariant metric. When choosing $g_{\mathfrak{k}} := (1 + \varepsilon^{-1})(g_0)|_{\mathfrak{k}}$, then (7.4) induces the metric g_0 on G and hence a normal homogeneous metric on G/H.

Recall from Theorem 7.3.2 that the metric $g_{\mathfrak{g}}$ from (7.5) has nonnegative curvature for some $\varepsilon > 0$ if and only if the semi-simple part of $\mathfrak{k}$ is an ideal of $\mathfrak{g}$. In particular, this is satisfied if $\dim(V) \leq 2$ in which case $\mathfrak{k}$ is at most one-dimensional and hence abelian. Therefore, we obtain the following result.

Theorem 7.5.4 *([5]) Let $M = G \times_K V$ be a homogeneous disk bundle of rank $\dim V \leq 2$. Then M admits a nonnegatively curved normal homogeneous collar metric.*

However, if $\dim V \geq 3$ and the bundle $M \to G/K$ is not essentially trivial, then the semi-simple part of $\mathfrak{k}$ is not an ideal. Thus, by Theorem 7.3.2, the left-invariant metric $g_{\mathfrak{g}}$ from (7.5) has some negative curvature on G. But in order for the submersion metric on $M = G \times_K V$ to be nonnegatively curved, it suffices that $g_{\mathfrak{g}}$ has nonnegative curvature on all planes *which are contained in* $\mathfrak{p}$. Some interesting examples where this occurs have been described in [9]. Once again, the proof of this theorem involves a direct calculation of the curvature formula for homogeneous metrics.

Theorem 7.5.5 *([9]) Let $H \subset K \subset G$ be compact Lie groups with Lie algebras $\mathfrak{h} \subset \mathfrak{k} \subset \mathfrak{g}$, and consider the decompositions from (7.1). If there exists $C > 0$ such that for all $X = X^{\mathfrak{m}} + X^{\mathfrak{s}}, Y = Y^{\mathfrak{m}} + Y^{\mathfrak{s}} \in \mathfrak{m} \oplus \mathfrak{s} = \mathfrak{p}$ we have*

$$|X^{\mathfrak{m}} \wedge Y^{\mathfrak{m}}| \leq C \cdot |[X, Y]|, \qquad (7.6)$$

then any left-invariant metric on G sufficiently close to g_0 which is Ad_H-invariant and is a constant multiple of g_0 on $\mathfrak{s}$ and $\mathfrak{h}$ (but arbitrary on $\mathfrak{m}$) has nonnegative sectional curvature on all planes contained in $\mathfrak{p}$.

That is, for all Lie groups $H \subset K \subset G$ satisfying this criterion, the metric $g_{\mathfrak{g}}$ from (7.5) has nonnegative curvature on all planes contained in $\mathfrak{p}$ for small $\varepsilon > 0$, hence if $K/H = S^V$, then the corresponding homogeneous vector bundle $G \times_K V \to G/K$ has a nonnegatively curved normal homogeneous collar metric.

As it turns out, condition (7.6) is almost necessary and sufficient for a homogeneous disk bundle to admit such a metric. Namely, we have

Theorem 7.5.6 *([10]) Let $G \times_K V \to G/K$ be a homogeneous vector bundle which admits a nonnegatively curved normal homogeneous collar metric, let $H \subset K$ be the subgroup for which $K/H = S^V$, and decompose the Lie algebras $\mathfrak{h} \subset \mathfrak{k} \subset \mathfrak{g}$ as in (7.1).*

If $\mathfrak{m}_1 \subset \mathfrak{m}$ is any non-trivial Ad_H-irreducible subspace such that $\mathfrak{m}$ contains no other Ad_H-irreducible factor equivalent to $\mathfrak{m}_1$, then there exists a constant $C > 0$ such that, for all $X = X^{\mathfrak{m}} + X^{\mathfrak{s}}, Y = Y^{\mathfrak{m}} + Y^{\mathfrak{s}} \in \mathfrak{m}_1 \oplus \mathfrak{s}$, condition (7.6) holds.

When K/H is isotropy irreducible, the choice $\mathfrak{m}_1 = \mathfrak{m}$ yields a converse to Theorem 7.5.5. The proof of this theorem uses Perelman's result on the existence of parallel vector fields on a totally convex set ([7]). This implies that along a normal geodesic $c(t)$ the metric on the principal orbit through $c(t)$ is induced by a metric of the form $g_t = (g_0)|_{\mathfrak{s}} + \tilde{g}_t|_{\mathfrak{k}}$. Then one shows that such a metric must have some negative curvature close to the collar unless (7.6) is satisfied.

With these two results, one can give an almost complete classification of homogeneous disk bundles which admit nonnegatively curved normal homogeneous collar metrics.

Theorem 7.5.7 *([10]) Let $M := G \times_K V \to G/K$ be a cohomogeneity one homogeneous vector bundle which is not essentially trivial. If M admits a G-invariant normal homogeneous collar metric of nonnegative curvature, then the rank of this bundle must be in the set $\{2, 3, 4, 6, 8\}$.*

By Theorem 7.5.4, all rank-two bundles admit a G-invariant normal homogeneous collar metric of nonnegative curvature. In the higher rank case, the situation is much more restricted. For rank eight, we have the following complete classification.

Theorem 7.5.8 *([10]) Let $M := G \times_K V \to G/K$ be a G-irreducible coho-mogeneity one homogeneous vector bundle which is not essentially trivial and such that $\dim_{\mathbb{R}} V = 8$. Then M admits a G-invariant normal homogeneous col-lar metric of nonnegative curvature if and only if M is finitely G-equivariantly covered by one of the following:*

(i) *$Spin(p + 9) \times_{Spin(8)} \mathbb{R}^8$ for $p \in \{0, 1, 2\}$, where $Spin(8)$ acts on $\mathbb{R}^8$ by a spin representation, and $Spin(8) \subset Spin(p + 9)$ is the lift of the standard inclusion $SO(8) \subset SO(p + 9)$.*

(ii) *$Spin(p + 8) \times_{Spin(7)} \mathbb{R}^8$ for $p \in \{0, 1\}$, where $Spin(7)$ acts on $\mathbb{R}^8$ by the spin representation, and $Spin(7) \subset Spin(p + 8)$ is the lift of the standard inclusion $SO(7) \subset SO(p + 8)$.*

(iii) *$Spin(7) \times_{Spin(6)} \mathbb{C}^4$, with the standard representation of $Spin(6) \cong SU(4)$ on $\mathbb{C}^4$.*

(iv) *A quotient of one of the preceding examples:*

    1 $(Spin(p + 9) \cdot G') \times_{Spin(8) \cdot H'} \mathbb{R}^8$ for $p \in \{1, 2\}$ and an arbitrary com-pact Lie group G' and $H' \subset Spin(p + 1) \cdot G'$ with $H' \not\subset G'$, which acts trivially on $\mathbb{R}^8$.

    2 $(Spin(9) \cdot G') \times_{Spin(7) \cdot H'} \mathbb{R}^8$ for an arbitrary compact Lie group G' and $H' \subset Spin(2) \cdot G' = S^1 \cdot G'$ with $H' \not\subset G'$, which acts trivially on $\mathbb{R}^8$.

    3 $(Spin(7) \cdot G') \times_{Spin(6) \cdot S^1 \cdot H'} \mathbb{C}^4$ for an arbitrary compact Lie group $G' \supset S^1 \cdot H'$, where $S^1 \subset G'$ acts on $\mathbb{C}^4$ by multiples of the identity, and $H' \not\subset G'$ acts trivially.

Here, for Lie groups L_1, L_2, we denote by $L_1 \cdot L_2$ the quotient of $L_1 \times L_2$ by a finite subgroup of the center. The term *G-irreducible* means that M is not G-equivariantly finitely covered by a bundle of the form $(G_1/H_1) \times M'$ with $\dim(G_1/H_1) > 0$ and M' a cohomogeneity one homogeneous vector bundle. This hypothesis is natural because, for G-reducible bundles, our problem easily reduces to deciding whether M' admits such a metric.

For rank 6 bundles, we obtain the following

Theorem 7.5.9 *([10]) Let $M := G \times_K V \to G/K$ be a G-irreducible coho-mogeneity one homogeneous vector bundle which is not essentially trivial and such that $\dim_{\mathbb{R}} V = 6$. Then M admits a G-invariant normal homogeneous col-lar metric of nonnegative curvature if and only if M is finitely G-equivariantly covered by one of the following:*

(i) *$SU(5) \times_{SU(4)} \mathbb{R}^6$, with the irreducible action of $SU(4) \cong Spin(6)$ on $\mathbb{R}^6$.*

(ii) *$(SU(5) \cdot G') \times_{SU(4) \cdot H'} \mathbb{R}^6$ for an arbitrary compact Lie group G' and $H' \subset S^1 \cdot G'$ with $S^1 \subset SU(5)$ being the centralizer of $SU(4)$, and $H' \not\subset G'$ acts trivially on $\mathbb{R}^6$.*

To describe our results for rank three and four bundles, we require some notation for subgroups of the exceptional Lie group G_2. Let $SO(4) \subset G_2$ and $SU(3) \subset G_2$ be the isotropy groups of the symmetric space $G_2/SO(4)$ and the sphere $S^6 = G_2/SU(3)$, respectively. After conjugating these groups appropriately, their intersection can be made isomorphic to $U(2)$, and we let $SU(2)_1 \subset SO(4) \cap SU(3) \subset G_2$ be the simple part of this intersection. Note that $SU(2)_1 \subset SO(4)$ is normal, and we denote its centralizer in G_2 by $SU(2)_3 \subset SO(4) \subset G_2$. (The subscripts of the $SU(2)$-subgroups of $SO(4)$ denote their maximal weight for the isotropy representation of $G_2/SO(4)$.) Using this notation, we can make the following statement about the rank-three case.

Theorem 7.5.10 *([10]) Let $M := G \times_K V$ be a G-irreducible cohomogeneity one homogeneous vector bundle which is not essentially trivial such that $\dim_{\mathbb{R}} V = 3$. If M admits a nonnegatively curved G-invariant normal homogeneous collar metric, then M must be finitely G-equivariantly covered by one of the following.*

(i) $M_1 = G_2 \times_{SO(4)} \mathfrak{su}(2)_3$, *where $SO(4)$ acts on $\mathfrak{su}(2)_3 \lhd \mathfrak{so}(4)$ by the adjoint representation.*

(ii) $M_2 = (Sp(p+1) \cdot G') \times_{Sp(1) \cdot H'} \mathfrak{sp}(1)$ *with* $H' \subset Sp(p) \cdot G'$, *where $Sp(1) \cdot H'$ acts on $\mathfrak{sp}(1) \lhd \mathfrak{sp}(1) \oplus \mathfrak{h}'$ by the adjoint representation.*

Further, M_1 admits such a metric, as does M_2 with $G' = 1$ and $H' = Sp(p)$.

Finally, in the rank-four case, we have the following examples, which are all related to those in Theorem 7.5.10.

Theorem 7.5.11 *([10]) The following cohomogeneity one homogeneous vector bundles (orbifold bundles, respectively) of rank four admit G-invariant normal homogeneous collar metrics of nonnegative curvature:*

(i) $G_2 \times_{SO(4)} (\mathbb{H}/\pm 1)$, *where $SU(2)_1 \subset SO(4)$ acts trivially and $SU(2)_3 \subset SO(4)$ by left multiplication on $\mathbb{H}/\pm 1$. Note that this is an orbifold bundle only.*

(ii) $(G_2 \times G') \times_{SO(4) \times SU(2)'} (\mathbb{H}/\pm 1)$, *where $SU(2)_1 \subset SO(4)$ acts trivially and $SU(2)_3 \subset SO(4)$ by left multiplication, whereas $SU(2)' \subset G'$ acts by right multiplication on $\mathbb{H}/\pm 1$ with G' arbitrary. Note that these are orbifold bundles only.*

(iii) $Sp(p+1) \times_{Sp(1) \times Sp(p)} \mathbb{H}$ *where $Sp(p)$ acts trivially and $Sp(1)$ by left multiplication on $\mathbb{H}$.*

(iv) $(Sp(p+1) \times G') \times_{Sp(1) \times Sp(p) \times Sp(1)'} \mathbb{H}$ *where* $Sp(p)$ *acts trivially and*
     $Sp(1)$ *by left multiplication, whereas* $Sp(1)' \subset G'$ *acts by right multi-*
     plication on $\mathbb{H}$ *with* G' *arbitrary.*

7.6 Cohomogeneity one manifolds

A *cohomogeneity one manifold* is a manifold M with an action of a compact
Lie group G such that $\dim(M/G) = 1$. For a cohomogeneity one manifold,
there are four possibilities for the quotient space.

(i) $M/G \cong \mathbb{R}$. In this case, M is equivariantly diffeomorphic to $\mathbb{R} \times G/H$
    for some subgroup $H \subset G$. In particular, M admits a G-invariant metric
    of nonnegative curvature.
(ii) $M/G \cong [0, \infty)$. In this case, M has one non-principal orbit and is a
    homogeneous vector bundle as described in the preceding section. In
    particular, M admits a G-invariant metric of nonnegative curvature.
(iii) $M/G \cong S^1$. Then M is finitely covered by $S^1 \times G/H$ for some subgroup
    $H \subset G$. Again, M admits a G-invariant metric of nonnegative curvature.
(iv) $M/G \cong [-1, 1]$. In this case, M has two non-principal orbits and can be
    written as

$$M = D_- \cup_{G/H} D_+,$$

where $D_\pm = G \times_{K_\pm} V_\pm$ are homogeneous disk bundles with common
boundary G/H. That is, $H \subset K_\pm$ and $K_\pm/H$ is the unit sphere $S^{V_\pm} \subset V_\pm$.

It is the last of these classes in which we are interested, as it is obtained by
glueing two homogeneous disk bundles together along their common boundary
which is a principal orbit. Conversely, given Lie groups $H \subset \{K_+, K_-\} \subset G$
such that $K_\pm/H$ is a sphere, then there is a unique cohomogeneity one manifold
which is determined by these groups. Thus, these groups can be used to identify
the manifold.

As it turns out, not all cohomogeneity one manifolds with two singular orbits
admit invariant metrics of nonnegative curvature ([3]). However, we have the
following result.

Theorem 7.6.1 *([11]) Let M be a closed cohomogeneity one manifold. Then
M admits G-invariant metrics of almost nonnegative curvature, i.e., there is a
sequence of G-invariant metrics g_n on M such that $diam(M, g_n)^2 sec(M, g_n) >
-1/n$.*

This is proven once again by using Cheeger deformations: Let g_M be any G-invariant metric on M, and let g_0 be a biinvariant metric on G. Then the metric g_λ on M induced by the submersion $(G \times M, \lambda g_0 + g_M) \to (M, g_\lambda)$ has uniformly bounded diameter for small $\lambda > 0$, and one verifies that the lower curvature bound of g_λ tends to 0 as $\lambda \to 0$.

From the results in the preceding section, we see that we obtain a nonnegatively curved metric on a cohomogeneity one manifold with two singular orbits if the corresponding two disk bundles admits nonnegatively curved normal homogeneous collar metrics. Thus, Theorem 7.5.4 immediately implies

Theorem 7.6.2 *([5]) Let M be a cohomogeneity one manifold with two non-principal orbits of codimension at most 2. Then M admits a G-invariant metric of nonnegative curvature.*

This result yields some spectacular examples of manifolds which admit nonnegatively curved metrics, such as all four (unoriented) diffeomorphism types of $\mathbb{RP}^5$ and ten out of the fourteen (unoriented) exotic spheres of dimension 7.

Unfortunately, the remaining homogeneous disk bundles admitting nonnegatively curved normal homogeneous collar metrics do not yield new examples of nonnegatively curved cohomogeneity one manifolds, since on glueing two of them, we obtain either the restriction of a homogeneous action on M, or there is a Riemannian submersion $\hat{M} \to M$, where $\hat{M}$ is known to have a nonnegatively curved metric.

For instance, by glueing together two disk bundles of the second type of Theorem 7.5.8 with $p = 0$, after applying outer automorphisms of $Spin(8)$, we conclude that the primitive cohomogeneity one manifold given by the group diagram $G_2 \subset \{Spin_+(7), Spin_-(7)\} \subset Spin(8)$ admits a metric of nonnegative curvature with a totally geodesic normal homogeneous principal orbit. However, this manifold is diffeomorphic to the sphere S^{15} ([4]).

References

[1] J. Cheeger, *Some examples of manifolds of nonnegative curvature*, J. Diff. Geom. 8 (1973) 623–628

[2] J. Cheeger, D. Gromoll, *On the structure of complete manifolds of nonnegative curvature*, Ann. of Math. 96 (1972) 413–443.

[3] K. Grove, L. Verdiani, B. Wilking, W. Ziller, *Nonnegative curvature obstructions in cohomogeneity one and the Kervaire spheres*, Ann. Sc. Norm. Super. Pisa, Cl. Sci. (5) 5, No. 2 (2006) 159–170

[4] K. Grove, B. Wilking, W. Ziller, *Positively curved cohomogeneity one manifolds and 3-Sasakian geometry*, J. Diff. Geom. 78 (2008) 33–111.

[5] K. Grove, W. Ziller, *Curvature and symmetry of Milnor spheres*, Ann. of Math. (2) 152 (2000) 331–367.

[6] L. Guijarro, *Improving the metric in an open manifold with nonnegative curvature*, Proc. Amer. Math. Soc., 126 (1998) 1541–1545

[7] G. Perelman, *Proof of the soul conjecture of Cheeger and Gromoll*, J. Diff. Geom. **40** (1994), 209–212.

[8] L. Schwachhöfer, *A remark on left invariant metrics on compact Lie groups*, Arch.Math. **90** (2008) 158–162

[9] L. Schwachhöfer, K. Tapp, *Homogeneous Metrics with nonnegative curvature*, Jour. Geom. Anal. 19 (2009) 929–943

[10] L. Schwachhöfer, K. Tapp, *Cohomogeneity one disk bundles with normal homogeneous collars*, Proc. London Math. Soc. 99, (2009) 609–632

[11] L. Schwachhöfer and W. Tuschmann, *Almost nonnegative curvature and cohomogeneity one*, Preprint no. 62/2001, Max-Planck-Institut für Mathematik in den Naturwissenschaften Leipzig, http://www.mis.mpg.de/cgi-bin/preprints.pl `

[12] W. Ziller, *Examples of Riemannian manifolds with nonnegative sectional curvature*, Metric and Comparison Geometry, Surv. Diff. Geom. 11, ed. K. Grove and J. Cheeger, Intern. Press, 2007

Author's address:

Fakultät für Mathematik,
Technische Universität Dortmund,
Vogelpothsweg 87,
44221 Dortmund,
Germany
lschwach@math.uni-dortmund.de

8

Morse theory and stable pairs

RICHARD A. WENTWORTH AND GRAEME WILKIN

Abstract

We study the Morse theory of the Yang-Mills-Higgs functional on the space of pairs (A, Φ), where A is a unitary connection on a rank 2 hermitian vector bundle over a compact Riemann surface, and Φ is a holomorphic section of (E, d_A''). We prove that a certain explicitly defined substratification of the Morse stratification is perfect in the sense of $\mathcal{G}$-equivariant cohomology, where $\mathcal{G}$ denotes the unitary gauge group. As a consequence, Kirwan surjectivity holds for pairs. It also follows that the twist embedding into higher degree induces a surjection on equivariant cohomology. This may be interpreted as a rank 2 version of the analogous statement for symmetric products of Riemann surfaces. Finally, we compute the $\mathcal{G}$-equivariant Poincaré polynomial of the space of τ-semistable pairs. In particular, we recover an earlier result of Thaddeus. The analysis provides an interpretation of the Thaddeus flips in terms of a variation of Morse functions.

8.1 Introduction

In this paper we revisit the notion of a stable pair on a Riemann surface. We introduce new techniques for the computation of the equivariant cohomology of moduli spaces. The main ingredient is a version of Morse theory in the spirit of Atiyah and Bott [1] adapted to the *singular* infinite dimensional space of holomorphic pairs.

Recall first the basic idea. Let E be a hermitian vector bundle over a closed Riemann surface M of genus $g \geq 2$. The space $\mathcal{A}(E)$ of unitary connections on E is an infinite dimensional affine space with an action of the group $\mathcal{G}$ of unitary gauge transformations. Via the Chern connection there is an isomorphism $A \mapsto d_A''$ between $\mathcal{A}(E)$ and the space of (integrable) Dolbeault operators (i.e.

holomorphic structures) on E. One of the key observations of Atiyah-Bott is that the Morse theory of a suitable $\mathcal{G}$-invariant functional on $\mathcal{A}(E)$, namely the Yang-Mills functional, gives rise to a smooth stratification (see also [6]). Moreover, this stratification is $\mathcal{G}$-equivariantly perfect in the sense that the long exact sequences for the equivariant cohomology of successive pairs split. Since $\mathcal{A}(E)$ is contractible, this gives an effective method, inductive on the rank of E, for computing the equivariant cohomology of the minimum, which consists of projectively flat connections.

Consider now a configuration space $\mathcal{B}(E)$ consisting of pairs (A, Φ), where $A \in \mathcal{A}(E)$ and Φ is a section of a vector bundle associated to E. We impose the condition that Φ be d_A''-holomorphic. Note that $\mathcal{B}(E)$ is still contractible, since an equivariant retraction of $\mathcal{B}(E)$ to $\mathcal{A}(E)$ is given by simply scaling Φ. It is therefore reasonable to attempt an inductive computation of equivariant cohomology as above. A problem arises, however, from the singularities caused by jumps in the dimension of the kernel as A varies. Nevertheless, the methods introduced in [8] for the case of Higgs bundles demonstrate that in certain cases this difficulty can be managed.

Below we apply this approach to the moduli space of rank 2, degree d, τ-semistable pairs $\mathfrak{M}_{\tau,d} = \mathcal{B}_{ss}^\tau(E) /\!\!/ \mathcal{G}^{\mathbb{C}}$ introduced by Bradlow [3] and Bradlow-Daskalopoulos [4]. In this case, Φ is a holomorphic section of E, and the Yang-Mills functional YM(A) is replaced by the Yang-Mills-Higgs functional YMH(A, Φ). We give a description of the algebraic and Morse theoretic stratifications of $\mathcal{B}(E)$. These stratifications, as well as the moduli space, depend on a real parameter τ, and since $\mathfrak{M}_{\tau,d}$ is nonempty only for $d/2 \leq \tau \leq d$, we shall always assume this bound for τ. For generic τ, $\mathcal{G}$ acts freely, and the quotient is geometric.

We will see that, as in [6, 7, 8], the algebraic and Morse stratifications agree (see Theorem 8.3.9). Because of singularities, however, the Morse stratification actually fails to be perfect in this case. We identify precisely how this comes about, and in fact we will show that this "failure of perfection" exactly cancels between different strata, so that there is a substratification that is indeed perfect (see Theorem 8.3.11). We formulate this result as

Theorem 8.1.1 (Equivariantly perfect stratification) *For every τ, $d/2 < \tau < d$, there is a $\mathcal{G}$-invariant stratification of $\mathcal{B}(E)$ defined via the Yang-Mills-Higgs flow that is perfect in $\mathcal{G}$-equivariant cohomology.*

The fact that perfection fails for the Morse stratification but holds for a substratification seems to be a new phenomenon. In any case, as with vector bundles, Theorem 8.1.1 allows us to compute the $\mathcal{G}$-equivariant cohomology of the open stratum $\mathcal{B}_{ss}^{\tau}(E)$. Explicit formulas in terms of symmetric products of M are given in Theorems 8.4.1 and 8.4.2.

There is a natural map (called the *Kirwan map*) from the cohomology of the classifying space $B\mathcal{G}$ of $\mathcal{G}$ to the equivariant cohomology of the stratum of τ-semistable pairs $\mathcal{B}_{ss}^{\tau}(E) \subset \mathcal{B}(E)$, coming from inclusion (see [13]). One of the consequences of the work of Atiyah-Bott is that the analogous map is surjective for the case of semistable bundles. The same is true for pairs:

Theorem 8.1.2 (Kirwan surjectivity) *The Kirwan map* $H^*(B\mathcal{G}) \to H_{\mathcal{G}}^*(\mathcal{B}_{ss}^{\tau}(E))$ *is surjective. In particular, for generic* τ, $H^*(B\mathcal{G}) \to H^*(\mathfrak{M}_{\tau,d})$ *is surjective.*

As noted above, for non-integer values of τ, $d/2 < \tau < d$, $\mathfrak{M}_{\tau,d}$ is a smooth projective algebraic manifold of dimension $d + 2g - 2$, and the equivariant cohomology of $\mathcal{B}_{ss}^{\tau}(E)$ is identical to the ordinary cohomology of $\mathfrak{M}_{\tau,d}$. The computation of equivariant cohomology presented here then recovers the result of Thaddeus in [20], who computed the cohomology using different methods. Namely, he gives an explicit description of the modifications, or "flips", in $\mathfrak{M}_{\tau,d}$ as the parameter τ varies. At integer values there is a change in stability conditions. Below, we show how the change in cohomology arising from a flip may be reinterpreted as a variation of the Morse function. This is perhaps not surprising in view of the construction in [5]. However, here we work directly on the infinite dimensional space. The basic idea is that there is a one parameter choice of Morse functions f_τ on $\mathcal{B}$. The minimum $f_\tau^{-1}(0)/\mathcal{G} \simeq \mathfrak{M}_{\tau,d}$, and the cohomology of $\mathfrak{M}_{\tau,d}$ may, in principle, be computed from the cohomology of the higher critical sets. As τ varies past certain critical values, new critical sets are created while others merge. Moreover, indices of critical sets can jump. All this taken together accounts for the change in topology of the minimum.

There are several important points in this interpretation. One is that the subvarieties responsible for the change in cohomology observed by Thaddeus as the parameter varies are somehow directly built into the Morse theory, even for a fixed τ, in the guise of higher critical sets. This example also exhibits computations at critical strata that can be carried out in the presence of singular normal *cusps*, as opposed to the singular normal vector bundles in [8]. These ideas may be useful for computations in higher rank or for other moduli spaces.

The critical set corresponding to minimal Yang-Mills connections, regarded as a subset of $\mathcal{B}(E)$ by setting $\Phi \equiv 0$, is special from the point of view of the Morse theory. In particular, essentially because of issues regarding Brill-Noether loci in the moduli space of vector bundles, we can only directly prove the perfection of the stratification at this step, and the crucial Morse-Bott lemma (Theorem 8.3.18), for $d > 4g - 4$. This we do in Section 8.3.5. By contrast, for the other critical strata there is no such requirement on the degree. Using this fact, we then give an inductive argument by twisting E by a positive line bundle and embedding $\mathcal{B}(E)$ into the space of pairs for higher degree, thus indirectly concluding the splitting of the associated long exact sequence even at minimal Yang-Mills connections in low degree (see Section 8.3.7).

This line of reasoning leads to another interesting consequence. For τ close to $d/2$, there is a surjective holomorphic map from $\mathfrak{M}_{\tau,d}$ to the moduli space of semistable rank 2 bundles of degree d. This is the rank 2 version of the Abel-Jacobi map [4]. In this sense, $\mathfrak{M}_{\tau,d}$ is a generalization of the d-th symmetric product $S^d M$ of M. Choosing an effective divisor on M of degree k, there is a natural inclusion $S^d M \hookrightarrow S^{d+k} M$, and it was shown by MacDonald in (14.3) of [16] that this inclusion induces a surjection on rational cohomology. A similar construction works for rank 2 pairs, except now $d \mapsto d + 2k$, while there is also a shift in the parameter $\tau \mapsto \tau + k$. We will prove the following.

Theorem 8.1.3 (Embedding in higher degree) *Let* $\deg E = d$ *and* $\deg \widetilde{E} = d + 2k$. *Then for all* $d/2 < \tau < d$, *the inclusion* $\mathcal{B}_{ss}^{\tau}(E) \hookrightarrow \mathcal{B}_{ss}^{\tau+k}(\widetilde{E})$ *described above induces a surjection on rational* $\mathcal{G}$-*equivariant cohomology. In particular, for generic* τ, *the inclusion* $\mathfrak{M}_{\tau,d} \hookrightarrow \mathfrak{M}_{\tau+k,d+2k}$ *induces a surjection on rational cohomology.*

Remark 8.1.4 It is also possible to construct a moduli space of pairs for which the isomorphism class of $\det E$ is fixed, indeed this is the space studied by Thaddeus in [20]. The explicit calculations in this paper are all done for the non-fixed determinant case, however it is worth pointing out here that the idea is essentially the same for the fixed determinant case, and that the only major difference between the two cases is in the topology of the critical sets. In particular, the indexing set $\Delta_{\tau,d}$ for the stratification is the same in both cases, and Theorems 8.1.1 and 8.1.2 hold for the fixed determinant spaces as well.

Acknowledgements The authors warmly thank the organizers of the Leeds conference, and especially Roger Bielawski, for putting together this volume. Thanks also to George Daskalopoulos for many discussions, and to the MPIM-Bonn, where some of the work on this paper was completed.

R.W. supported in part by NSF grant DMS-1037094

8.2 Stable pairs

8.2.1 The Harder-Narasimhan stratification

Throughout this paper, E will denote a rank 2 hermitian vector bundle on M of positive degree $d = \deg E$. We will regard E as a smooth complex vector bundle, and when endowed with a holomorphic structure that is understood, we will use the same notation for the holomorphic bundle.

Recall that a holomorphic bundle E of degree d is *stable* (resp. *semistable*) if $\deg L < d/2$ (resp. $\deg L \leq d/2$) for all holomorphic line subbundles $L \subset E$.

Definition 8.2.1 For a stable holomorphic bundle E, set $\mu_+(E) = d/2$. For E unstable, let

$$\mu_+(E) = \sup\{\deg L : L \subset E \text{ a holomorphic line subbundle}\}$$

For a holomorphic section $\Phi \not\equiv 0$ of E, define $\deg \Phi$ to be the number of zeros of Φ, counted with multiplicity. Finally, for a holomorphic pair (E, Φ) let

$$\mu_-(E, \Phi) = \begin{cases} d - \deg \Phi & \Phi \not\equiv 0 \\ d - \mu_+(E) & \Phi \equiv 0 \end{cases}$$

Definition 8.2.2 ([3]) Given τ, a holomorphic pair (E, Φ) is called τ-*stable* (resp. τ-*semistable*) if

$$\mu_+(E) < \tau < \mu_-(E, \Phi) \qquad (\text{resp. } \mu_+(E) \leq \tau \leq \mu_-(E, \Phi))$$

As with holomorphic bundles, there is a notion of s-equivalence of strictly semistable objects. The set $\mathfrak{M}_{\tau,d}$ of isomorphism classes of semistable pairs, modulo s-equivalence, has the structure of a projective variety. Note that $\mathfrak{M}_{\tau,d}$ is empty if $\tau \notin [d/2, d]$. For non-integer values of $\tau \in (d/2, d)$, semistable is equivalent to stable, and $\mathfrak{M}_{\tau,d}$ is smooth.

Let $\mathcal{A} = \mathcal{A}(E)$ denote the infinite dimensional affine space of holomorphic structures on E, $\mathcal{G}$ the group of unitary gauge transformations, and $\mathcal{G}^{\mathbb{C}}$ its complexification. The space $\mathcal{A}$ may be identified with Dolbeault operators $A \mapsto d_A'' : \Omega^0(E) \to \Omega^{0,1}(E)$, with the inverse of d_A'' given by the Chern connection with respect to the fixed hermitian structure. When we want to emphasize the holomorphic bundle, we write (E, d_A'').

$$\mathcal{B} = \mathcal{B}(E) = \left\{(A, \Phi) \in \mathcal{A} \times \Omega^0(E) : d_A'' \Phi = 0\right\} \tag{8.1}$$

Let

$$\mathcal{B}_{ss}^\tau = \left\{(A, \Phi) \in \mathcal{B} : ((E, d_A''), \Phi) \text{ is } \tau\text{-semistable}\right\}$$

Then $\mathfrak{M}_{\tau,d} = \mathcal{B}_{ss}^{\tau} /\!\!/ \mathcal{G}^{\mathbb{C}}$, where the double slash identifies s-equivalent orbits. For generic values of τ, semistability implies stability and $\mathcal{G}$ acts freely, and so this is a geometric quotient.

We now describe the stratification of $\mathcal{B}$ associated to the Harder-Narasimhan filtration, which has an important relationship to the Morse theory picture that will be discussed below in Section 8.3.2. In the case of rank 2 bundles, this stratification is particularly easy to describe. For convenience, throughout this section we fix a generic τ, $d/2 < \tau < d$ (it suffices to assume $4\tau \notin \mathbb{Z}$). Genericity is used only to give a simple description of the strata in terms of δ. The extension to special values of τ is straightforward (see Remark 8.2.11).

Note that stability of the pair fails if either of the inequalities in Definition 8.2.2 fails. The two inequalities are not quite independent, but there are some cases where only one fails and others where both fail. If the latter, it seems natural to filter by the *most destabilizing of the two*. With this in mind, we make the following.

Definition 8.2.3 For a holomorphic pair (E, Φ), let

$$\delta(E, \Phi) = \max\{\tau - \mu_-(E, \Phi), \mu_+(E) - \tau, 0\}$$

Note that δ takes on a discrete and infinite set of nonnegative real values, and is upper semi-continuous, since both μ_+ and $-\mu_-$ are (observe that $\deg \Phi \leq \mu_+(E)$). We denote the ordered set of such δ by $\Delta_{\tau,d}$. Clearly, δ is an integer modulo $\pm\tau$, or $\delta = \tau - d/2$. Because of the genericity of τ, the former two possibilities are mutually exclusive:

Lemma 8.2.4 *There is a disjoint union* $\Delta_{\tau,d} \setminus \{0\} = \Delta_{\tau,d}^+ \cup \Delta_{\tau,d}^-$, *with*

$$\delta \in \Delta_{\tau,d}^+ \iff \delta = \tau - \mu_-(E, \Phi), \text{ for some pair } (E, \Phi)$$
$$\delta \in \Delta_{\tau,d}^- \iff \delta = \mu_+(E) - \tau, \text{ for some pair } (E, \Phi)$$

Lemma 8.2.5 *Suppose* $(E, \Phi) \notin \mathcal{B}_{ss}^{\tau}$, $\Phi \not\equiv 0$. *Then*

(i) if $\deg \Phi \geq d/2$, $\delta(E, \Phi) = \mu_+(E) - d + \tau$.
(ii) if $d - \tau \leq \deg \Phi < d/2$, $\delta(E, \Phi) = \deg \Phi - d + \tau$.
(iii) if $0 \leq \deg \Phi < d - \tau$, $\delta(E, \Phi) = \mu_+(E) - \tau$.

If $\Phi \equiv 0$, *then* $\delta(E, \Phi) = \mu_+(E) - d + \tau$.

Proof If $\deg \Phi \geq d/2$, then the line subbundle generated by Φ is the maximal destabilizing subbundle of E. Hence, $\mu_+(E) = \deg \Phi$, $\mu_-(E, \Phi) = d - \mu_+(E)$, and so (i) follows from the fact that $\tau > d/2$. For (ii), consider the extension $0 \to L_1 \to E \to L_2 \to 0$, where $\Phi \in H^0(L_1)$. Then $\deg L_2 = d - \deg \Phi$, so $\mu_+(E, \Phi) \leq d - \deg \Phi$. It follows that $\mu_+(E) - \tau \leq 0$. For part (iii), $0 \leq \deg \Phi < d - \tau$ implies $\tau < \mu_-(E, \Phi)$. The last statement is clear, since $\tau > d/2$ implies $\tau - \mu_-(E, \Phi) = \mu_+(E) - d + \tau > \mu_+(E) - \tau$. $\qquad \square$

Corollary 8.2.6 $\Delta^-_{\tau,d} \subset (0, d - \tau]$.

Proof Indeed, if (E, Φ) is unstable and $\delta(E, \Phi) = \mu_+(E) - \tau$, then by (3) it must be that E is unstable and $\Phi \not\equiv 0$. From the Harder-Narasimhan filtration (cf. [14]) $0 \to L_2 \to E \to L_1 \to 0$, the projection of Φ to L_1 must also be nonzero, since $\deg \Phi < \deg L_2$. Hence, $\deg L_1 = d - \mu_+(E) \geq 0$, and so $d - \tau \geq \delta(E, \Phi)$. $\qquad \square$

Remark 8.2.7 If $\delta \in \Delta^+_{\tau,d}$ and $\delta < \tau - d/2$, then $\delta \leq \tau - d/2 - 1/2$. Indeed, if $\delta + d - \tau = k \in \mathbb{Z}$, the condition forces $k < d/2$; hence, $k \leq d/2 - 1/2$.

Let $I_{\tau,d} = [\tau - d/2, 2\tau - d)$. We are ready to describe the τ-Harder-Narasimhan stratification. First, for $j > d/2$, let $\mathcal{A}_j \subset \mathcal{A}$ be the set of holomorphic bundles E of Harder-Narasimhan type $\mu_+(E) = j$. We also set $\mathcal{A}_{d/2} = \mathcal{A}_{ss}$. There is an obvious inclusion $\mathcal{A}_j \subset \mathcal{B} : A \mapsto (A, 0)$.

(0) $\delta = 0$: The open stratum $\mathcal{B}^\tau_0 = \mathcal{B}^\tau_{ss}$ consists of τ-semistable pairs.

(I_a) $\delta \in \Delta^+_{\tau,d} \cap I_{\tau,d}$: Then we include the strata $\mathcal{A}_{\delta+d-\tau}$. Note that this includes the semistable stratum $\mathcal{A}_{ss}$. The bundles in this strata that are not semistable have a unique description as extensions

$$0 \longrightarrow L_1 \longrightarrow E \longrightarrow L_2 \longrightarrow 0 \qquad\qquad (8.2)$$

where $\deg L_1 = \mu_+(E) = \delta + d - \tau$.

(I_b) $\delta \in \Delta^+_{\tau,d} \cap [2\tau - d, +\infty)$: Then $\mathcal{B}^\tau_\delta = \{(E, \Phi) : \mu_+(E) = \delta + d - \tau\}$. These are extensions (8.2), $\deg L_1 = \mu_+(E) = \delta + d - \tau$, $\Phi \subset H^0(L_1)$.

(II^+) $\delta \in \Delta^+_{\tau,d} \cap (0, 2\tau - d)$: Then $\mathcal{B}^\tau_\delta = \{(E, \Phi) : \deg \Phi = \delta + d - \tau\}$. These are extensions (8.2), $\deg L_1 = \delta + d - \tau$, $\Phi \subset H^0(L_1)$.

(II^-) $\delta \in \Delta^-_{\tau,d}$: Then $\mathcal{B}^\tau_\delta = \{(E, \Phi) : \mu_+(E) = \delta + \tau, \ \deg \Phi < d/2\}$. These are extensions

$$0 \longrightarrow L_2 \longrightarrow E \longrightarrow L_1 \longrightarrow 0$$

where $\deg L_2 = \mu_+(E)$, and the projection of Φ to $H^0(L_1)$ is nonzero.

For simplicity of notation, when τ is fixed we will mostly omit the superscript: $\mathcal{B}_\delta = \mathcal{B}^\tau_\delta$.

Remark 8.2.8 It is simple to verify that the stratification obtained above coincides with the possible Harder-Narasimhan filtrations of pairs (E, Φ) considered as *coherent systems* (see [15, 18, 12]).

It will be convenient to organize $\Delta_{\tau,d}$ by the slope of the subbundle in the maximal destabilizing subpair. Define $j : \Delta_{\tau,d} \setminus \{0\} \to \{d/2\} \cup \{k \in \mathbb{Z} : k \geq d - \tau\}$ by

$$j(\delta) = \begin{cases} \delta + d - \tau, & \delta \in \Delta_{\tau,d}^+ \\ \delta + \tau, & \delta \in \Delta_{\tau,d}^- \end{cases} \tag{8.3}$$

Notice that $j(\delta) = \deg L_1$ for $\delta \in \Delta_{\tau,d}^+$, and $j(\delta) = \deg L_2$ for $\delta \in \Delta_{\tau,d}^-$, where L_1, L_2 refer to the line subbundles of E in the filtrations above. Note that j is surjective. It is precisely 2-1 on the image of $\Delta_{\tau,d}^-$ and 1-1 elsewhere (if d odd; otherwise $d/2$ labels both the stratum $\mathcal{A}_{ss}$ and the strictly semistable bundles of type $\mathbf{II}^+$). It is not order preserving but is, of course, order preserving on each of $\Delta_{\tau,d}^\pm$ separately.

Definition 8.2.9 For $\delta \in \Delta_{\tau,d}$, let

$$X_\delta = \bigcup_{\delta' \leq \delta, \, \delta' \in \Delta_{\tau,d}} \mathcal{B}_{\delta'} \cup \bigcup_{\delta' \leq \delta, \, \delta' \in \Delta_{\tau,d}^+ \cap I_{\tau,d}} \mathcal{A}_{j(\delta')}$$

For $\delta \in \Delta_{\tau,d}^+ \cap I_{\tau,d}$, let

$$X_\delta' = \bigcup_{\delta' \leq \delta, \, \delta' \in \Delta_{\tau,d}} \mathcal{B}_{\delta'} \cup \bigcup_{\delta' < \delta, \, \delta' \in \Delta_{\tau,d}^+ \cap I_{\tau,d}} \mathcal{A}_{j(\delta')}$$

For $\delta \notin \Delta_{\tau,d}^+ \cap I_{\tau,d}$, let

$$X_\delta' = \bigcup_{\delta' < \delta, \, \delta' \in \Delta_{\tau,d}} \mathcal{B}_{\delta'} \cup \bigcup_{\delta' < \delta, \, \delta' \in \Delta_{\tau,d}^+ \cap I_{\tau,d}} \mathcal{A}_{j(\delta')}$$

We call the collection $\{X_\delta, X_\delta'\}_{\delta \in \Delta_{\tau,d}}$ the τ-*Harder-Narasimhan stratification* of $\mathcal{B}$.

Note that $X_{\delta_1} \subset X_\delta' \subsetneq X_\delta \subset X_{\delta_2}'$, where δ_1 is the predecessor and δ_2 is the successor of δ in $\Delta_{\tau,d}$. If $\delta \notin \Delta_{\tau,d}^+ \cap I_{\tau,d}$, then $X_\delta' = X_{\delta_1}$ and $X_\delta = X_{\delta_2}'$. In the special case $\delta = \tau - d/2$, we have

$$X_{\tau - d/2} = X_{\tau-d/2}' \cup \mathcal{A}_{ss} \tag{8.4}$$

$$X_{\tau-d/2}' = \begin{cases} X_{\delta_1} & \text{if } d \text{ is odd} \\ X_{\delta_1} \cup \mathcal{B}_{\tau - d/2} & \text{if } d \text{ is even} \end{cases} \tag{8.5}$$

The following is clear.

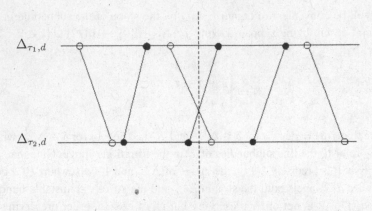

Figure 8.1 A "flip"

Proposition 8.2.10 *The sets* $\{X_\delta, X'_\delta\}_{\delta \in \Delta_{\tau,d}}$, *are locally closed in* $\mathcal{B}$, $\mathcal{G}$-*invariant, and satisfy*

$$\mathcal{B} = \bigcup_{\delta \in \Delta_{\tau,d}} X_\delta = \bigcup_{\delta \in \Delta_{\tau,d}} X'_\delta$$

$$\overline{\mathcal{B}}_\delta \subset \bigcup_{\delta \le \delta', \delta' \in \Delta_{\tau,d}} \mathcal{B}_{\delta'} = \mathcal{B}_\delta \cup \bigcup_{\delta < \delta', \delta' \in \Delta_{\tau,d}} \mathcal{B}'_{\delta'}$$

$$\overline{\mathcal{B}}'_\delta \subset \bigcup_{\delta \le \delta', \delta' \in \Delta_{\tau,d}} \mathcal{B}'_{\delta'} = \mathcal{B}'_\delta \cup \bigcup_{\delta < \delta', \delta' \in \Delta_{\tau,d}} \mathcal{B}_{\delta'}$$

Remark 8.2.11 To extend this stratification in the case of nongeneric τ, we define the sets $\Delta^\pm_{\tau,d}$ and the corresponding strata as above. For $\delta \in \Delta^+_{\tau,d} \cap \Delta^-_{\tau,d}$ there are two or possibly three components with the same label.

Let us note the following behavior as τ varies. For $\tau_1 \le \tau_2$, there is a well-defined map $\Delta_{\tau_1,d} \to \Delta_{\tau_2,d}$ given by $\delta \mapsto \max\{\delta \pm (\tau_2 - \tau_1), 0\}$, where $\pm$ depends on $\delta \in \Delta^\pm_{\tau,d}$. Hence, elements of $\Delta^+_{\tau,d}$ (white circles in Figure 8.1 below) "move" to the right, and elements of $\Delta^-_{\tau,d}$ (dark circles) "move" to the left as τ increases. The map is an order preserving bijection *provided* τ_1, τ_2 are in a connected component of $(d/2, d) \setminus C_d$, where

$$C_d = \{\tau_c \in (d/2, d) : 2\tau_c \in \mathbb{Z} \text{ if } d \text{ even}, 4\tau_c \in \mathbb{Z} \text{ if } d \text{ odd}\} \quad (8.6)$$

However, as τ_2 crosses an element of C_d, there is a "flip" in the stratification. When this flip occurs at $\delta = 0$, this is the phenomenon discovered by Thaddeus [20]; the discussion here is an extension of this effect to the entire stratification.

Finally, we will also have need to refer to the Harder-Narasimhan stratification of the space $\mathcal{A}$ of unitary connections on E. We denote this by

$$X_j^{\mathcal{A}} = \bigcup_{d/2 \le j' \le j} \mathcal{A}_{j'} \tag{8.7}$$

The following statement will be used later on. It is an immediate consequence of the descriptions of the strata above.

Lemma 8.2.12 *Consider the projection* $\mathrm{pr} : \mathcal{B} \to \mathcal{A}$. *Then*

$$\mathrm{pr}(\mathcal{B}_\delta) = \mathcal{A}_{j(\delta)}, \quad \delta \in \Delta_{\tau,d}^- \cup \left(\Delta_{\tau,d}^+ \cap [\tau - d/2, +\infty) \right)$$
$$\mathrm{pr}(\mathcal{B}_\delta) = X_{d-j(\delta)}^{\mathcal{A}}, \quad \delta \in \Delta_{\tau,d}^+ \cap (0, \tau - d/2)$$

8.2.2 Deformation theory

Fix a conformal metric on M, normalized[1] for convenience so that $\mathrm{vol}(M) = 2\pi$. Infinitesimal deformations of $(A, \Phi) \in \mathcal{B}$ modulo equivalence are described by the following elliptic complex, which we denote by $\mathcal{C}_{(A,\Phi)}$ (cf. [4]).

$$\mathcal{C}_{(A,\Phi)}^0 \xrightarrow{\ D_1\ } \mathcal{C}_{(A,\Phi)}^1 \xrightarrow{\ D_2\ } \mathcal{C}_{(A,\Phi)}^2$$

$$\Omega^0(\mathrm{End}\, E) \xrightarrow{\ D_1\ } \Omega^{0,1}(\mathrm{End}\, E) \oplus \Omega^0(E) \xrightarrow{\ D_2\ } \Omega^{0,1}(E) \tag{8.8}$$

$$D_1(u) = (-d_A'' u, u\Phi), \quad D_2(a, \varphi) = d_A'' \varphi + a\Phi$$

Here, D_1 is the linearization of the action of the complex gauge group $\mathcal{G}^{\mathbb{C}}$ on $\mathcal{B}$, and D_2 is the linearization of the condition $d_A'' \Phi = 0$. Note that $D_2 D_1 = 0$ if $(A, \Phi) \in \mathcal{B}$. The hermitian metric gives adjoint operators

$$D_1^*(a, \varphi) = -(d_A'')^* a + \varphi \Phi^*, \quad D_2^*(\beta) = (\beta \Phi^*, (d_A'')^* \beta) \tag{8.9}$$

The spaces of *harmonic forms* are by definition

$$\mathcal{H}^0(\mathcal{C}_{(A,\Phi)}) = \ker D_1$$
$$\mathcal{H}^1(\mathcal{C}_{(A,\Phi)}) = \ker D_1^* \cap \ker D_2$$
$$\mathcal{H}^2(\mathcal{C}_{(A,\Phi)}) = \ker D_2^*$$

Vectors in $\Omega^{0,1}(\mathrm{End}\, E) \oplus \Omega^0(E)$ that are orthogonal to the $\mathcal{G}^{\mathbb{C}}$-orbit through (A, Φ) are in $\ker D_1^*$, and a *slice* for the action of $\mathcal{G}^{\mathbb{C}}$ on $\mathcal{B}$ is therefore given by

$$\mathcal{S}_{(A,\Phi)} = \ker D_1^* \cap \left\{ (a, \varphi) \in \Omega^{0,1}(\mathrm{End}\, E) \oplus \Omega^0(E) : D_2(a, \varphi) + a\varphi = 0 \right\} \tag{8.10}$$

---

[1] More generally, the scale invariant parameter is $\tau \, \mathrm{vol}(M)/2\pi$.

Define the *slice map*

$$\Sigma : (\ker D_1)^{\perp} \times \mathcal{S}_{(A,\Phi)} \to \mathcal{B}$$
$$(u, a, \varphi) \mapsto e^u \cdot (A + a, \Phi + \varphi) \tag{8.11}$$

The proof of the following may be modeled on [21, Proposition 4.12]. We omit the details.

Proposition 8.2.13 *The slice map Σ is a local homeomorphism from a neighborhood of 0 in $(\ker D_1)^{\perp} \times \mathcal{S}_{(A,\Phi)}$ to a neighborhood of (A, Φ) in $\mathcal{B}$.*

The *Kuranishi map* is defined by

$$\Omega^{0,1}(\text{End } E) \oplus \Omega^0(E) \overset{k}{\longrightarrow} \Omega^{0,1}(\text{End } E) \oplus \Omega^0(E)$$
$$k(a, \varphi) = (a, \varphi) + D_2^* \circ G_2(a\varphi)$$

where G_2 denotes the Green's operator associated to the laplacian $D_2(D_2)^*$. We have the following standard result (cf. [14, Chapter VII] for the case of holomorphic bundles over a Kähler manifold and [4] for this case).

Proposition 8.2.14 *The Kuranishi map k maps $\mathcal{S}_{(A,\Phi)}$ to harmonics $\mathcal{H}^1(\mathcal{C}_{(A,\Phi)})$, and in a neighborhood of zero it is a local homeomorphism onto its image. Moreover, if $\mathcal{H}^2(\mathcal{C}_{(A,\Phi)}) = \{0\}$, then k is a local homeomorphism $\mathcal{S}_{(A,\Phi)} \to \mathcal{H}^1(\mathcal{C}_{(A,\Phi)})$.*

The following is immediate from (8.8) and (8.9).

Lemma 8.2.15 *Given $(A, \Phi) \in \mathcal{B}$, if $\Phi \neq 0$ then $\mathcal{H}^0(\mathcal{C}_{(A,\Phi)}) = \mathcal{H}^2(\mathcal{C}_{(A,\Phi)}) = \{0\}$. If $H^1(E) = \{0\}$ then $\mathcal{H}^2(\mathcal{C}_{(A,\Phi)}) = \{0\}$.*

We will be interested in the deformation complex along higher critical sets of the Yang-Mills-Higgs functional. As we will see in the next section, in addition to the Yang-Mills connections (where $\Phi \equiv 0$), the other critical sets correspond to split bundles $E = L_1 \oplus L_2$, $(A, \Phi) = (A_1 \oplus A_2, \Phi_1 \oplus \{0\})$, with $\deg L_1 = j \geq \deg L_2 = d - j$. Here, $j = j(\delta)$ for some $\delta \in \Delta_{\tau,d}^+$, or $j = d - j(\delta)$ for some $\delta \in \Delta_{\tau,d}^-$. The set of all such critical points will therefore be denoted by $\eta_\delta \subset \mathcal{B}$. We will denote the components of $\text{End } E \simeq L_i \otimes L_j^*$ in the complex by $u_{ij}, a_{ij}, \varphi_{ij}$.

In this case, $\mathcal{H}^1(\mathcal{C}_{(A,\Phi)})$ consists of all (a, φ) satisfying

$$(d_A'')^* a_{12} = 0 \qquad (d'')^* a_{22} = 0$$
$$(d'')^* a_{11} - \varphi_1 \Phi_1^* = 0 \quad (d_A'')^* a_{21} - \varphi_2 \Phi_1^* = 0 \tag{8.12}$$
$$d_{A_1}'' \varphi_1 + a_{11} \Phi_1 = 0 \quad d_{A_2}'' \varphi_2 + a_{21} \Phi_1 = 0$$

We use this formalism to define deformation retractions in a neighborhood of $(A, \Phi) \in \mathcal{B}$ in two cases. First, we have

Lemma 8.2.16 *Suppose* $(A, \Phi) = (A_1 \oplus A_2, \Phi_1 \oplus 0)$ *is a split pair as above,* $\Phi_1 \neq 0$. *Let*

$$S^{neg.}_{(A,\Phi)} = \{(a, \varphi) \in S_{(A,\Phi)} : a_{ij} = 0, (ij) \neq (21), \text{ and } \varphi_1 = 0\}$$
$$S'_{(A,\Phi)} = \{(a, \varphi) \in S_{(A,\Phi)} : (a_{21}, \varphi_2) \neq 0\}$$

Then there is an equivariant deformation retraction $S^{neg.}_{(A,\Phi)} \hookrightarrow S_{(A,\Phi)}$ *which restricts to a deformation retraction* $S^{neg.}_{(A,\Phi)} \setminus \{0\} \hookrightarrow S'_{(A,\Phi)}$.

Proof By Lemma 8.2.15 and Proposition 8.2.14, the Kuranishi map gives a homeomorphism of the slice with $\mathcal{H}^1(\mathcal{C}_{(A,\Phi)})$. Hence, it suffices to define the retraction there. For this we take

$$r_t(a_{11}, a_{12}, a_{21}, a_{22}; \varphi_1, \varphi_2) = (ta_{11}, ta_{12}, a_{21}, ta_{22}; t\varphi_1, \varphi_2), \ t \in [0, 1]$$

Notice that this preserves the equations in (8.12). $\qquad\qquad\square$

Second, near minimal Yang-Mills connections, we find a similar retraction under the assumption that $\mathcal{H}^2(\mathcal{C}_{(A,\Phi)})$ vanishes.

Lemma 8.2.17 *Suppose* $d > 4g - 4$ *and* Λ *is semistable. Let*

$$S^{neg.}_{(A,0)} = \{(a, \varphi) \in S_{(A,0)} : a = 0\}$$
$$S'_{(A,0)} = \{(a, \varphi) \in S_{(A,0)} : \varphi \neq 0\}$$

Then there is an equivariant deformation retraction $S^{neg.}_{(A,0)} \hookrightarrow S_{(A,0)}$ *which restricts to a deformation retraction* $S^{neg.}_{(A,0)} \setminus \{0\} \hookrightarrow S'_{(A,0)}$.

Proof Let E be the holomorphic bundle given by A. Since E is semistable, so is $E^* \otimes K_M$, where K_M is the canonical bundle of M. On the other hand, by the assumption, $\deg(E^* \otimes K_M) = 4g - 4 - \deg E < 0$. Hence, by Serre duality, $H^1(E) \simeq H^0(E^* \otimes K_M)^* = \{0\}$. Given a, let $\mathcal{H}_a$ denote harmonic projection to $\ker d''_{A+a}$. It follows that for a in a small neighborhood of the origin in the slice, $\mathcal{H}_a$ is a continuous family. We can therefore define the deformation retraction explicitly by

$$r_t(a, \varphi) = (ta, \mathcal{H}_{ta}(\varphi)), \ t \in [0, 1]$$

For a sufficiently small neighborhood of the origin in the slice, this preserves the set $S'_{(A,0)}$. It is also clearly equivariant. $\qquad\qquad\square$

8.3 Morse theory

8.3.1 The τ-vortex equations

Let $\mu(A, \Phi) = *F_A - i\Phi\Phi^*$. Then $*\mu$ is a moment map for the action of $\mathcal{G}$ on $\mathcal{B} \subset \mathcal{A} \times \Omega^0(E)$. Let $\tau > 0$ be a positive parameter and define the Yang-Mills-Higgs functional

$$f_\tau(A, \Phi) = \|\mu + i\tau \cdot \mathrm{id}\|^2 \qquad (8.1)$$

Solutions to the τ-vortex equations are the absolute minima of f_τ:

$$\mu(A, \Phi) + i\tau \cdot \mathrm{id} = 0 \qquad (8.2)$$

Theorem 8.3.1 (Bradlow [3])

$$\mathfrak{M}_{\tau,d} = \{(A, \Phi) \in \mathcal{B} : \mu(A, \Phi) + i\tau \cdot \mathrm{id} = 0\} / \mathcal{G}.$$

If the space of solutions to the τ-vortex equations is nonempty, then τ must satisfy the following restriction.

$$\mu + i\tau \cdot \mathrm{id} = *F_A - i\Phi\Phi^* + i\tau \cdot \mathrm{id} = 0$$
$$\implies \quad \frac{i}{2\pi} \int_M \mathrm{Tr}(*F_A - i\Phi\Phi^*) = 2\tau \iff \deg E + \|\Phi\|^2 = 2\tau \qquad (8.3)$$

Therefore $2\tau \geq d$ (with strict inequality if we want to ensure that $\Phi \neq 0$). Theorem 2.1.6 of [3] shows that a solution to the τ-vortex equations which is not τ-stable must split. Moreover, since $\mathrm{rank} E = 2$ the solutions can only split if τ is an integer. In particular, for a generic choice of τ solutions to (8.2) must be τ-stable. In general, critical sets of f_τ can be characterized in terms of a decomposition of the holomorphic structure of E. The critical point equations for the functional f_τ are

$$d_A''(\mu + i\tau \cdot \mathrm{id}) = 0 \qquad (8.4)$$
$$(\mu + i\tau \cdot \mathrm{id})\Phi = 0 \qquad (8.5)$$

There are three different types of critical points.

- **(0)** Absolute minimum $f_\tau^{-1}(0)$.
- **(I)** Yang-Mills connections with $\Phi = 0$. Then either A is an irreducible Yang-Mills minimum or E splits holomorphically as $E = L_1 \oplus L_2$. The latter exist for all values of $\deg L_1 \geq d/2$ and the existence of the critical points is independent of the choice of τ. However, as shown below the Morse index does depend on τ. If E is semistable (resp. $\deg L_1 < \tau$) we call this a critical point of type $\mathbf{I}_a$, and we label it $\delta = \tau - d/2$

(resp. $\delta = \deg L_1 - d + \tau$). If $\deg L_1 > \tau$ it is of type $\mathbf{I}_b$, and set $\delta = \deg L_1 - d + \tau$.

(**II**) E splits holomorphically as $E = L_1 \oplus L_2$, and $\Phi \in H^0(L_1) \setminus \{0\}$. On L_1 we have

$$*F_{A_1} - i\Phi\Phi^* = -i\tau, \quad \|\Phi\|^2 = 2\pi(\tau - \deg L_1)$$

Therefore $\deg L_1 < \tau$. Further subdivide these depending upon $\deg L_1$.

(**II$^-$**) $\deg L_1 \leq d - \tau, \delta = d - \deg L_1 - \tau$;

(**II$^+$**) $d - \tau < \deg L_1 < \tau, \delta = \deg L_1 - d + \tau$;

Let $S^d M$ denote the d-th symmetric product of the Riemann surface M, and $J_d(M)$ the Jacobian variety of degree d line bundles on M. For future reference we record the following

Proposition 8.3.2 *For $\delta \in \Delta_{\tau,d} \setminus \{0\}$,*

$$H_{\mathcal{G}}^*(\eta_\delta) =$$

$$\begin{cases} H_{\mathcal{G}}^*(\mathcal{A}_{ss}) & \text{\textit{Type} } \mathbf{I}, \delta = \tau - d/2 \\ H^*(J_{j(\delta)}(M) \times J_{d-j(\delta)}(M)) \otimes H^*(BU(1) \times BU(1)) & \text{\textit{Type} } \mathbf{I}, \delta \neq \tau - d/2 \\ H^*(S^{j(\delta)} M \times J_{d-j(\delta)}(M)) \otimes H^*(BU(1)) & \text{\textit{Type} } \mathbf{II}^+ \\ H^*(S^{d-j(\delta)} M \times J_{j(\delta)}(M)) \otimes H^*(BU(1)) & \text{\textit{Type} } \mathbf{II}^- \end{cases}$$

8.3.2 The gradient flow

Consider the negative gradient flow of the Yang-Mills-Higgs functional f_τ defined on the space $\mathcal{B} \subset \mathcal{A} \times \Omega^0(E)$. Since the functional is very similar to that studied in [10], we only sketch the details of the existence and convergence of the flow and focus on showing that the Morse stratification induced by the flow is equivalent to the Harder-Narasimhan stratification described in Section 8.2.1.

The gradient flow equations are

$$\frac{\partial A}{\partial t} = 2 * d_A(\mu + i\tau), \quad \frac{\partial \Phi}{\partial t} = -4i(\mu + i\tau)\Phi \tag{8.6}$$

Theorem 8.3.3 *The gradient flow of f_τ with initial conditions in $\mathcal{B}$ exists for all time and converges to a critical point of f_τ in the smooth topology.*

A standard calculation (cf. [3, Section 4]) shows that f_τ can be re-written as

$$f_\tau = \int_X \left(|F_A|^2 + |d'_A \Phi|^2 + |\Phi\Phi^*|^2 - 2\tau |\Phi|^2 + |\tau|^2 \right) dvol + 4\tau \deg E \tag{8.7}$$

This is very similar to the functional YMH studied in [10], and the proof for existence of the flow for all positive time follows the same structure (which is in turn modeled on Donaldson's proof for the Yang-Mills functional in [9]), therefore we omit the details. An important part of the proof worth mentioning here is that the flow is generated by the action of $\mathcal{G}^{\mathbb{C}}$, i.e. for all $t \in [0, \infty)$ there exists $g(t) \in \mathcal{G}^{\mathbb{C}}$ such that the solution $(A(t), \Phi(t))$ to the flow equations (8.6) with initial condition (A, Φ) is given by $(A(t), \Phi(t)) = g(t) \cdot (A, \Phi)$.

To show that the gradient flow converges, one can use the results of Theorem B of [11] (where again, the functional is not exactly the same as f_τ, but it has the same structure and so the proof of convergence is similar). The statement of [11, Theorem B] only describes smooth convergence along a subsequence (since they also study the higher dimensional case where bubbling occurs), and to extend this to show that the limit is unique we use the Lojasiewicz inequality technique of [19] and [17]. The key estimate is contained in the following proposition.

Proposition 8.3.4 *Let* (A_∞, Φ_∞) *be a critical point of* f_τ. *Then there exist* $\varepsilon_1 > 0$, *a positive constant* C, *and* $\theta \in \left(0, \frac{1}{2}\right)$, *such that*

$$\|(A, \Phi) - (A_\infty, \Phi_\infty)\| < \varepsilon_1$$

implies that

$$\|\nabla f_\tau(A, \Phi)\|_{L^2} \geq C \, |f_\tau(A, \Phi) - f_\tau(A_\infty, \Phi_\infty)|^{1-\theta} \qquad (8.8)$$

The proof is similar to that in [21], and so is omitted.

The rest of the proof of convergence then follows the analysis in [21] for Higgs bundles. The key result is the following proposition, which is the analog of [21, Proposition 3.7] (see also [19] or [17, Proposition 7.4]).

Proposition 8.3.5 *Each critical point* (A, Φ) *of* f_τ *has a neighborhood* U *such that if* $(A(t), \Phi(t))$ *is a solution of the gradient flow equations for* f_τ *and* $(A(T), \Phi(T)) \in U$ *for some* T, *then either* $f_\tau(A(t), \Phi(t)) < f_\tau(A, \Phi)$ *for some* t, *or* $(A(t), \Phi(t))$ *converges to a critical point* (A', Φ') *such that* $f_\tau(A', \Phi') = f_\tau(A, \Phi)$. *Moreover, there exists* ε *(depending on* U*) such that* $\|(A', \Phi') - (A, \Phi)\| < \varepsilon$.

The next step is the main result of this section: The Morse stratification induced by the gradient flow of f_τ is the same as the τ-Harder–Narasimhan stratification described in Section 8.2.1. First recall the Hitchin–Kobayashi correspondence from Theorem 8.3.1, and the distance-decreasing result from [10], which can be re-stated as follows.

Lemma 8.3.6 (Hong [10]) *Let (A_1, Φ_1) and (A_2, Φ_2) be two pairs related by an element $g \in \mathcal{G}^{\mathbb{C}}$. Then the distance between the $\mathcal{G}$-orbits of $(A_1(t), \Phi_1(t))$ and $(A_2(t), \Phi_2(t))$ is non-increasing along the flow.*

Recall that the critical sets associated to each stratum are given in Section 8.3.1, and that the critical set associated to the stratum $\mathcal{B}_\delta$ is denoted η_δ. Define $S_\delta \subset \mathcal{B}$ to be the subset of pairs that converge to a point in C_δ under the gradient flow of f_τ. The next lemma gives some standard results about the critical sets of f_τ.

Lemma 8.3.7
  (i) The critical set η_δ is the minimum of the functional f_τ on the stratum $\mathcal{B}_\delta$.
  (ii) The closure of each $\mathcal{G}^{\mathbb{C}}$ orbit in $\mathcal{B}_\delta$ intersects the critical set η_δ.
  (iii) There exists $\varepsilon > 0$ (depending on τ) such that $(A, \Phi) \in \eta_\delta$ and $(A', \Phi') \in \eta_{\delta'}$ with $\delta \neq \delta'$ implies that $\|(A, \Phi) - (A', \Phi')\| \geq \varepsilon$.

Proof Since these results are analogous to standard results for the Yang–Mills functional (see for example [1], [6], or [7]), and the proof for holomorphic pairs is similar, we only sketch the idea of the proof here.

- The first statement follows by noting that the convexity of the norm-square function $\| \cdot \|^2$ shows that the minimum of f_τ on each extension class occurs at a critical point. This can be checked explicitly for each of the types $\mathbf{I}_a$, $\mathbf{I}_b$, $\mathbf{II}^+$, and $\mathbf{II}^-$.
- To see the second statement, simply scale the extension class and apply Theorem 8.3.1 (the Hitchin–Kobayashi correspondence) to the graded object of the filtration (cf. [7, Theorem 3.10] for the Yang–Mills case).
- The third statement can be checked by noting that (modulo the $\mathcal{G}$-action) the critical sets are compact, and then explicitly computing the distance between distinct critical sets. $\qquad\square$

As a consequence we have

Proposition 8.3.8
  (i) Each critical set η_δ has a neighborhood V_δ such that $V_\delta \cap \mathcal{B}_\delta \subset S_\delta$.
  (ii) $S_\delta \cap \mathcal{B}_\delta$ is $\mathcal{G}^{\mathbb{C}}$-invariant.

Proof Proposition 8.3.5 implies that there exists a neighborhood V_δ of each critical set η_δ such that if $(A, \Phi) \in V_\delta$ then the flow with initial conditions (A, Φ) either flows below η_δ, or converges to a critical point close to η_δ. Since f_τ is minimized on each Harder–Narasimhan stratum $\mathcal{B}_\delta$ by the critical set η_δ, the flow is generated by the action of $\mathcal{G}^{\mathbb{C}}$, and the strata $\mathcal{B}_\delta$ are $\mathcal{G}^{\mathbb{C}}$-invariant, then the first alternative cannot occur if $(A, \Phi) \in \mathcal{B}_\delta \cap V_\delta$. Since the critical sets are a finite distance apart, then (by shrinking V_δ if necessary) the limit must be contained in η_δ. Therefore $V_\delta \cap \mathcal{B}_\delta \subset S_\delta$, which completes the proof of (i).

To prove (ii), for each pair $(A, \Phi) \in S_\delta \cap \mathcal{B}_\delta$, let

$$Y_{(A,\Phi)} = \left\{ g \in \mathcal{G}^\mathbb{C} : g \cdot (A, \Phi) \in S_\delta \cap \mathcal{B}_\delta \right\}.$$

The aim is to show that $Y_{(A,\Phi)} = \mathcal{G}^\mathbb{C}$. Firstly we note that since the group Γ of components of $\mathcal{G}^\mathbb{C}$ is the same as that for the unitary gauge group $\mathcal{G}$, the flow equations (8.6) are $\mathcal{G}$-equivariant, and the critical sets η_δ are $\mathcal{G}$-invariant, then it is sufficient to consider the connected component of $\mathcal{G}^\mathbb{C}$ containing the identity. Therefore the problem reduces to showing that $Y_{(A,\Phi)}$ is open and closed. Openness follows from the continuity of the group action, the distance-decreasing result of Lemma 8.3.6, and the result in part (i). Closedness follows by taking a sequence of points $\{g_k\} \subset Y_{(A,\Phi)}$ that converges to some $g \in \mathcal{G}^\mathbb{C}$, and observing that the distance-decreasing result of Lemma 8.3.6 implies that the flow with initial conditions $g \cdot (A, \Phi)$ must converge to a limit close to the $\mathcal{G}$-orbit of the limit of the flow with initial conditions $g_k \cdot (A, \Phi)$ for some large k. Since the critical sets are $\mathcal{G}$-invariant, and critical sets of different types are a finite distance apart, then by taking k large enough (so that $g_k \cdot (A, \Phi)$ is close enough to $g \cdot (A, \Phi)$) we see that the limit of the flow with initial conditions $g \cdot (A, \Phi)$ must be in η_δ. Therefore $Y_{(A,\Phi)}$ is both open and closed. $\qquad\square$

Theorem 8.3.9 *The Morse stratification by gradient flow is the same as the Harder–Narasimhan stratification in Definition 8.2.9.*

Proof The goal is to show that $\mathcal{B}_\delta \subseteq S_\delta$ for each δ. Let $x \in \mathcal{B}_\delta$. By Lemma 8.3.7 (ii) the closure of the orbit $\mathcal{G}^\mathbb{C} \cdot x$ intersects η_δ, therefore there exists $g \in \mathcal{G}^\mathbb{C}$ such that $g \cdot x \in V_\delta \cap \mathcal{B}_\delta \subseteq S_\delta$ by Proposition 8.3.8 (i). Since $S_\delta \cap \mathcal{B}_\delta$ is $\mathcal{G}^\mathbb{C}$-invariant by Proposition 8.3.8 (ii), then $x \in \mathcal{B}_\delta \cap S_\delta$ also, and therefore $\mathcal{B}_\delta \subseteq S_\delta$. Since $\{\mathcal{B}_\delta\}$ and $\{S_\delta\}$ are both stratifications of $\mathcal{B}$, then we have $\mathcal{B}_\delta = S_\delta$ for all δ. $\qquad\square$

Remark 8.3.10 While we have identified the stable strata of the critical sets with the Harder-Narasimhan strata, the ordering on the set $\Delta_{\tau,d}$ coming from the values of YMH is more complicated. Since this will not affect the calculations, we continue to use the ordering already defined in Section 2.

We may now reformulate the main result, Theorem 8.1.1. The key idea is to define a substratification of $\{X_\delta, X'_\delta\}_{\delta \in \Delta_{\tau,d}}$ by combining $\mathcal{B}_\delta$ and $\mathcal{A}_{j(\delta)}$ for $\delta \in \Delta^+_{\tau,d} \cap I_{\tau,d}$. In other words, this is simply $\{X_\delta\}_{\delta \in \Delta_{\tau,d}}$. We call this the *modified* Morse stratification.

Theorem 8.3.11 *The modified Morse stratification $\{X_\delta\}_{\delta \in \Delta_{\tau,d}}$ is $\mathcal{G}$-equivariantly perfect in the following sense: For all $\delta \in \Delta_{\tau,d}$, the long exact sequence*

$$\cdots \longrightarrow H_{\mathcal{G}}^*(X_\delta, X_{\delta_1}) \longrightarrow H_{\mathcal{G}}^*(X_\delta) \longrightarrow H_{\mathcal{G}}^*(X_{\delta_1}) \longrightarrow \cdots \quad (8.9)$$

splits. Here, δ_1 denotes the predecessor of δ in $\Delta_{\tau,d}$.

8.3.3 Negative normal spaces

For critical points $(A, \Phi) \in \eta_\delta$, a tangent vector

$$(a, \varphi) \in \Omega^{0,1}(\text{End } E) \oplus \Omega^0(E)$$

is an eigenvector for the Hessian of f_τ if

$$i[\mu + i\tau \cdot \text{id}, a] = \lambda a \quad (8.10)$$

$$i(\mu + i\tau \cdot \text{id})\varphi = \lambda\varphi \quad (8.11)$$

Let $V_{(A,\Phi)}^{neg.} \subset \Omega^{0,1}(\text{End } E) \oplus \Omega^0(E)$ denote the span of all such (a, φ) with $\lambda < 0$. This is clearly $\mathcal{G}$-invariant, since f_τ is. Let $\mathcal{S}_{(A,\Phi)}$ be the slice at (A, Φ). Then we set $\nu_\delta \cap \mathcal{S}_{(A,\Phi)} = V_{(A,\Phi)}^{neg.} \cap \mathcal{S}_{(A,\Phi)}$. Using Proposition 8.2.13, this gives a well-defined $\mathcal{G}$-invariant subset $\nu_\delta \subset \mathcal{B}$, which we call the *negative normal space at η_δ*. By definition, η_δ is a closed subset of ν_δ.

We next describe ν_δ in detail for each of the critical sets:

(I_a) Recall that in this case $\Phi \equiv 0$. If E semistable, the negative eigenspace of the Hessian is $H^0(E)$. To see this, note that since $\Phi = 0$ then $i(\mu + i\tau \cdot \text{id}) = (d/2 - \tau) \cdot \text{id}$ is a negative constant multiple of the identity (by assumption $\tau > d/2$). Therefore $i[\mu + i\tau \cdot \text{id}, a] = 0$, and $a = 0$. Then the slice equations imply $\varphi \in H^0(E)$. If $E = L_1 \oplus L_2$, then $\mathcal{H}^2(\mathcal{C}_{(A,0)})$ is nonzero in general. From the slice equations, we see that the negative eigendirections ν_δ of the Hessian are given by

$$d_{A_2}''\varphi_2 + a_{21}\varphi_1 = 0, \ (a_{21}, \varphi_1) \in H^{0,1}(L_1^* L_2) \oplus H^0(L_1) \quad (8.12)$$

(I_b) This is similar to the case above, except now for negative directions, $\varphi_1 \equiv 0$. We therefore conclude that ν_δ is given by

$$H^{0,1}(L_1^* L_2) \oplus H^0(L_2) \quad (8.13)$$

Note that if $\delta > \tau$, then $\deg L_2 = d - j(\delta) < 0$, and so ν_δ^- has constant dimension $\dim_{\mathbb{C}} H^{0,1}(L_1^* L_2) = 2j(\delta) - d + g - 1$.

(II^+) In this case, $\Phi \not\equiv 0$, so by Lemma 8.2.15, $\mathcal{H}^2(\mathcal{C}_{(A,0)}) = 0$, and the slice is homeomorphic to $\mathcal{H}^1(\mathcal{C}_{(A,0)})$ via the Kuranishi map. The negative eigenspace of the Hessian is then just

$$(d_A'')^* a_{21} - \varphi_2 \Phi_1^* = 0, \quad d_A'' \varphi_2 + a_{21} \Phi_1 = 0 \tag{8.14}$$

(**II⁻**) This is similar to above, except now $\varphi_2 \equiv 0$. Hence, the fiber of ν_δ is given by

$$H^{0,1}(L_2^* L_1) \tag{8.15}$$

Note that $\dim_{\mathbb{C}} H^{0,1}(L_2^* L_1) = 2j(\delta) - d + g - 1$.

To see (**II⁺**) and (**II⁻**), we need to compute the solutions to (8.10) and (8.11), which involves knowing the value of $i(\mu + i\tau \cdot \mathrm{id})$ on the critical set. Equation (8.4) shows that

$$i(\mu + i\tau \cdot \mathrm{id}) = \begin{pmatrix} \lambda_1 & 0 \\ 0 & \lambda_2 \end{pmatrix}$$

where $\lambda_1 \in \Omega^0(L_1^* L_1)$ and $\lambda_2 \in \Omega^0(L_2^* L_2)$ are constant. Since $\Phi \in H^0(L_1) \setminus \{0\}$, then (8.5) shows that $\lambda_1 = 0$. Since λ_2 is constant, the integral over M becomes

$$\lambda_2 = \frac{1}{2\pi} \int_M \lambda_2 \, dvol = \frac{i}{2\pi} \int_M F_{A_2} - \frac{1}{2\pi} \int_M \tau \, dvol = \deg L_2 - \tau$$

Therefore, if $d - \tau < \deg L_1 = d - \deg L_2$, then $\deg L_2 < \tau$ and so λ_2 is negative. Similarly, if $\deg L_1 < d - \tau$ then λ_2 is positive. Equation (8.10) then shows that $a \in \Omega^{0,1}(L_1^* L_2)$ if $d - \tau < \deg L_1$, and $a \in \Omega^{0,1}(L_2^* L_1)$ if $\deg L_1 < d - \tau$. Similarly, if $d - \tau < \deg L_1$ then $\varphi \in \Omega^0(L_2)$, and if $\deg L_1 < d - \tau$ then $\varphi = 0$. Equations (8.14) and (8.15) then follow from the slice equations.

The following lemma describes the space of solutions to (8.12) when φ_1 is fixed.

Lemma 8.3.12 *Fix φ_1. When $\varphi_1 = 0$ then the space of solutions $\{(a_{21}, \varphi_2)\}$ to (8.12) is isomorphic to $H^{0,1}(L_1^* L_2) \oplus H^0(L_2)$. When $\varphi_1 \neq 0$ then the space of solutions $\{(a_{21}, \varphi_2)\}$ to (8.12) has dimension $\deg L_1$.*

Proof The first case (when $\varphi_1 = 0$) is easy, since the equations for $a \in \Omega^{0,1}(L_1^* L_2)$ and $\varphi_2 \in \Omega^0(L_2)$ become

$$d_A''^* a = 0, \quad d_A'' \varphi_2 = 0. \tag{8.16}$$

In the second case (when $\varphi_1 \neq 0$ is fixed), note (8.12) implies that $\mathcal{H}(a\varphi_1) = 0$, where $\mathcal{H}$ denotes the harmonic projection $\Omega^{0,1}(L_2) \to H^{0,1}(L_2)$. Hence, it suffices to show that the map

$$H^{0,1}(L_1^* L_2) \to H^{0,1}(L_2) \tag{8.17}$$

given by multiplication with φ_1 (followed by harmonic projection) is surjective. For then, since $\deg L_1^* L_2 < 0$, we have by Riemann–Roch that the dimension of (8.12) is $h^0(L_2) + h^1(L_1^* L_2) - h^1(L_2) = \deg L_1$. By Serre duality, (8.17) is surjective if and only if $H^0(KL_2^*) \to H^0(KL_2^* L_1)$ is injective. But since $\varphi_1 \neq 0$, multiplication gives an injection of sheaves $\mathcal{O} \hookrightarrow L_1$, and the result follows by tensoring and taking cohomology. $\qquad\qquad\square$

Lemma 8.3.13 *The space of solutions to* (8.14) *has constant dimension* $= \deg L_1 = j(\delta)$.

Proof Consider the subcomplex $\mathcal{C}_{(A,\Phi)}^{LT}$

$$\Omega^0(L_1^* L_2)) \xrightarrow{D_1} \Omega^{0,1}(L_1^* L_2) \oplus \Omega^0(L_2) \xrightarrow{D_2} \Omega^1(L_2) \qquad (8.18)$$

Since $\Phi \neq 0$, by Lemma 8.2.15 the cohomology at the ends of the complex (8.18) vanishes, and we have (by Riemann–Roch)

$$\begin{aligned}
\dim_{\mathbb{C}} \mathcal{H}^1(\mathcal{C}_{(A,\Phi)}^{LT}) &= \dim_{\mathbb{C}}(\ker D_1^* \cap \ker D_2) \\
&= h^1(L_1^* L_2) + h^0(L_2) - h^1(L_2) - h^0(L_1^* L_2) \\
&= -\deg L_1^* L_2 + g - 1 + \deg L_2 + (1 - g) \\
&= \deg L_1
\end{aligned}$$

$\qquad\square$

We summarize the the above considerations with

Corollary 8.3.14 *The fiber of ν_δ is linear of constant dimension for critical sets of type* $\mathbf{II}^\pm$, *and for those of type* $\mathbf{I}_b$ *provided* $\delta \notin \Delta_{\tau,d}^+ \cap [\tau - d/2, \tau]$. *The complex dimension of the fiber in these cases is* $\sigma(\delta)$, *where*

$$\sigma(\delta) = \begin{cases} 2j(\delta) - d + g - 1 & \text{if type } \mathbf{I}_b \text{ or } \mathbf{II}^- \\ j(\delta) & \text{if type } \mathbf{II}^+ \end{cases}$$

Remark 8.3.15 The strata for $\delta \in I_{\tau,d}$ have two components corresponding to the strata $\mathcal{A}_{j(\delta)}$ and $\mathcal{B}_\delta$. When there is a possible ambiguity, we will distinguish these by the notation $\nu_{I,\delta}$ for the negative normal spaces to strata of type $\mathbf{I}_a$ or $\mathbf{I}_b$, and $\nu_{II,\delta}$ for the negative normal spaces to strata of type $\mathbf{II}^+$ or $\mathbf{II}^-$.

8.3.4 Cohomology of the negative normal spaces

As in [8], at certain critical sets – namely, those of type $\mathbf{I}_a$, $\mathbf{I}_b$ where $\delta \in \Delta_{\tau,d}^+ \cap [\tau - d/2, \tau]$ – the negative normal directions are not necessarily constant in dimension. In the present case, they are not even linear. In order to carry out the computations, we appeal to a relative sequence by considering special subspaces with better behavior.

Definition 8.3.16 For $\delta \in \Delta_{\tau,d}^+ \cap (\tau - d/2, \tau]$, let $\nu_{I,\delta}$ be the negative normal space to a critical set with $\Phi \equiv 0$, as in Section 8.3.3. Define

$$\nu'_{I,\delta} = \{(a, \varphi_1, \varphi_2) \in \nu_{I,\delta} : (a, \varphi_1, \varphi_2) \neq 0\}$$
$$\nu''_{I,\delta} = \{(a, \varphi_1, \varphi_2) \in \nu_{I,\delta} : a \neq 0\}$$

The goal of this section is the proof of the following

Proposition 8.3.17

$$\delta \in \Delta_{\tau,d}^+ \cap (\tau - d/2, \tau] : H_{\mathcal{G}}^*(\nu_{I,\delta}, \nu''_{I,\delta}) \simeq H_{S^1 \times S^1}^{*-2(2j(\delta)-d+g-1)}(\eta_{j(\delta)}^{\mathcal{A}})$$

$$(8.19)$$

$$\delta \in \Delta_{\tau,d}^+ \cap (2\tau - d, \tau] : H_{\mathcal{G}}^*(\nu'_{I,\delta}, \nu''_{I,\delta})$$
$$\simeq H_{S^1}^{*-2(2j(\delta)-d+g-1)}(S^{d-j(\delta)}M \times J_{j(\delta)}(M))$$

$$(8.20)$$

$$\delta \in \Delta_{\tau,d}^+ \cap (\tau-d/2, 2\tau-d) : H_{\mathcal{G}}^*(\nu'_{I,\delta}, \nu''_{I,\delta}) \simeq H_{S^1}^{*-2j(\delta)}(S^{j(\delta)}M \times J_{d-j(\delta)}(M))$$

$$(8.21)$$

$$\oplus H_{S^1}^{*-2(2j(\delta)-d+g-1)}(S^{d-j(\delta)}M \times J_{j(\delta)}(M))$$

Proof Fix $E = L_1 \oplus L_2$. Consider first the case $\tau > \deg L_1 = j(\delta) > d/2$, and $\deg L_2 = d - j(\delta) < d/2$. Define the following spaces

$$\omega_\delta = \{(A_1, A_2, a, \varphi_1, \varphi_2) \in \nu_{I,\delta} : (a, \varphi_2) \neq 0\}$$
$$Z_\delta^- = \{(A_1, A_2, a, \varphi_1, \varphi_2) \in \nu_{I,\delta} : \varphi_1 = 0\}$$
$$Z_\delta' = \{(A_1, A_2, a, \varphi_1, \varphi_2) \in \nu_{I,\delta} : \varphi_1 = 0, (a, \varphi_2) \neq 0\}$$
$$Y_\delta' = \{(A_1, A_2, a, \varphi_1, \varphi_2) \in \nu_{I,\delta} : \varphi_1 \neq 0\}$$
$$Y_\delta'' = \{(A_1, A_2, a, \varphi_1, \varphi_2) \in \nu_{I,\delta} : \varphi_1 \neq 0, (a, \varphi_2) \neq 0\}$$
$$T_\delta = \{(A_1, A_2, a, \varphi_1, \varphi_2) \in \nu_{I,\delta} : \varphi_1 \neq 0, (a, \varphi_2) = 0\}$$

Note that $Y'_\delta = \nu_{I,\delta} \setminus Z^-_\delta = \nu'_{I,\delta} \setminus Z'_\delta$ and $Y''_\delta = \omega_\delta \setminus Z'_\delta$. Consider the following commutative diagram.

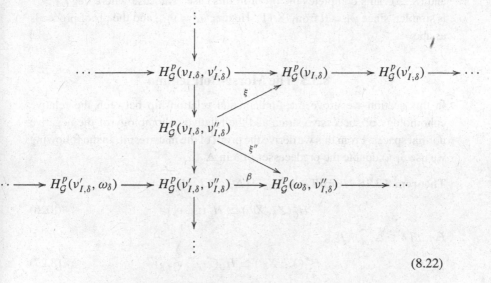

$$(8.22)$$

- First, it follows as in the proof of [8, Thm. 2.3] that the pair $(\nu_{I,\delta}, \nu''_{I,\delta})$ is homotopic to the Atiyah–Bott pair $(X^{\mathcal{A}}_{j(\delta)}, X^{\mathcal{A}}_{j(\delta)-1})$. Hence, (8.19) follows from [1].
- Consider the pair $(\nu'_{I,\delta}, \omega_\delta)$. Excision of Z'_δ gives the isomorphism

$$H^*_{\mathcal{G}}(\nu'_{I,\delta}, \omega_\delta) \cong H^*_{\mathcal{G}}(\nu'_{I,\delta} \setminus Z'_\delta, \omega_\delta \setminus Z'_\delta) \cong H^*_{\mathcal{G}}(Y'_\delta, Y''_\delta) \qquad (8.23)$$

The space $Y''_\delta = Y'_\delta \setminus T_\delta$, and Lemma 8.3.12 shows that Y'_δ is a vector bundle over T_δ with fibre dimension $= \deg L_1$. Therefore the Thom isomorphism implies

$$H^*_{\mathcal{G}}(Y'_\delta, Y''_\delta) = H^*_{\mathcal{G}}(Y'_\delta, Y'_\delta \setminus X'_\delta) \cong H^{*-2j(\delta)}_{\mathcal{G}}(T_\delta)$$

and therefore

$$H^*_{\mathcal{G}}(\nu'_{I,\delta}, \omega_\delta) = H^*_{\mathcal{G}}(Y'_\delta, Y''_\delta) = H^{*-2j(\delta)}_{S^1}(S^{j(\delta)} M \times J_{d-j(\delta)}(M)) \qquad (8.24)$$

- Consider $(\omega_\delta, \nu''_{I,\delta})$. By retraction, the pair is homotopic to the intersection with $\varphi_1 = 0$. It then follows exactly as in [8] (or the argument above) that

$$H^*_{\mathcal{G}}(\omega_\delta, \nu''_{I,\delta}) \cong H^{*-2(2j(\delta)-d+g-1)}_{S^1}(S^{d-j(\delta)} M \times J_{j(\delta)}(M)) \qquad (8.25)$$

(Recall that $\dim H^{0,1}(L^*_1 L_2) = 2j(\delta) - d + g - 1$ by Riemann–Roch, and that $\deg L_2 = d - j(\delta)$.)

It then follows as in [8] that ξ'', and hence also β, is surjective. This implies that the lower horizontal exact sequence splits, and (8.21) follows from (8.24) and (8.25). This completes the proof in this case. The case where $\deg L_1 > \tau$ is simpler, since $\varphi_1 \equiv 0$ from (8.11). Hence, $\omega_\delta = \nu'_{I,\delta}$, and the proof proceeds as above. $\square$

8.3.5 The Morse-Bott lemma

In this section we prove the fundamental relationship between the relative cohomology of successive strata and the relative cohomology of the negative normal spaces. From this we derive the proof of the main result. In the following we use δ_1 to denote the predecessor of δ in $\Delta_{\tau,d}$.

Theorem 8.3.18 *For all* $\delta \in \Delta_{\tau,d} \setminus \left(\Delta_{\tau,d}^+ \cap I_{\tau,d} \right)$,

$$H_{\mathcal{G}}^*(X_\delta, X_{\delta_1}) \simeq H_{\mathcal{G}}^*(\nu_\delta, \nu'_\delta) \tag{8.26}$$

For all $\delta \in \Delta_{\tau,d}^+ \cap I_{\tau,d}$,

$$H_{\mathcal{G}}^*(X'_\delta, X_{\delta_1}) \simeq H_{\mathcal{G}}^*(\nu_{II,\delta}, \nu'_{II,\delta}) \tag{8.27}$$

For all $\delta \in \Delta_{\tau,d}^+ \cap I_{\tau,d}$, $\delta \neq \tau - d/2$,

$$H_{\mathcal{G}}^*(X_\delta, X'_\delta) \simeq H_{\mathcal{G}}^*(\nu_{I,\delta}, \nu'_{I,\delta}) \tag{8.28}$$

Eq. (8.28) *also holds for* $\delta = \tau - d/2$, *provided* $d > 4g - 4$. *In the statements above,* δ_1 *denotes the predecessor of* δ *in* $\Delta_{\tau,d}$.

First, we give a proof of (8.26) in the case $\delta \notin \Delta_{\tau,d}^+ \cap [\tau - d/2, \tau]$. By excision and convergence of the gradient flow, there is a neighborhood U of η_δ such that

- U is $\mathcal{G}$-invariant;
- U is the union of images of slices $\mathcal{S}_{(A,\Phi)}$, where $(A, \Phi) \in \eta_\delta$;
- $H_{\mathcal{G}}^*(X_\delta, X_{\delta_1}) \simeq H_{\mathcal{G}}^*(U, U \setminus (U \cap \mathcal{B}_\delta))$

Notice that for each slice $\mathcal{S}_{(A,\Phi)} \cap U \setminus (U \cap \mathcal{B}_\delta) = \mathcal{S}'_{(A,\Phi)} \cap U$, where the latter is defined as in Lemma 8.2.16. By the lemma, it follows that the pair $(U, U \setminus (U \cap \mathcal{B}_\delta))$ locally retracts to $(\nu_\delta, \nu'_\delta)$. On the other hand, by Corollary 8.3.14, ν_δ is a bundle over η_δ. It follows by continuity as in [2], that there is a $\mathcal{G}$-equivariant retraction of the pair $(\nu_\delta, \nu'_\delta) \hookrightarrow (U, U \setminus (U \cap \mathcal{B}_\delta))$. The result therefore follows in this case. We also note that by Corollary 8.3.14 and the Thom isomorphism,

$$H_{\mathcal{G}}^*(\nu_\delta, \nu'_\delta) \simeq H_{\mathcal{G}}^{*-2\sigma(\delta)}(\eta_\delta) \tag{8.29}$$

Remark 8.3.19 Notice that by Corollary 8.3.14 the same argument also proves (8.27). For $d > 4g - 4$, we can use Lemma 8.2.17 in the same way to derive (8.28) for $\delta = \tau - d/2$. In this case, by the Thom isomorphism, we have

$$H_{\mathcal{G}}^*(X_{\tau-d/2}, X'_{\tau-d/2}) \simeq H_{\mathcal{G}}^{*-2(d+2-2g)}(\mathcal{A}_{ss}) \tag{8.30}$$

Lemma 8.3.20 *For* $\delta \notin \Delta_{\tau,d}^+ \cap [\tau - d/2, \tau]$, *or if* $\delta = \tau - d/2$ *and* $d > 4g - 4$, *then the long exact sequence* (8.9) *splits. Similarly, the long exact sequence*

$$\cdots \longrightarrow H_{\mathcal{G}}^p(X'_\delta, X_{\delta_1}) \longrightarrow H_{\mathcal{G}}^p(X'_\delta) \longrightarrow H_{\mathcal{G}}^p(X_{\delta_1}) \longrightarrow \cdots \tag{8.31}$$

splits for all $\delta \in I_{\tau,d}$.

Proof Indeed, since (8.26) holds in this case, we have

$$
\begin{array}{ccccccc}
\cdots \longrightarrow & H_{\mathcal{G}}^p(X_\delta, X_{\delta_1}) & \overset{\alpha}{\longrightarrow} & H_{\mathcal{G}}^p(X_\delta) & \longrightarrow & H_{\mathcal{G}}^p(X_{\delta_1}) & \longrightarrow \cdots \\
& \downarrow{\scriptstyle\cong} & & \downarrow & & & \\
& H_{\mathcal{G}}^p(\nu_\delta, \nu'_\delta) & \overset{\beta}{\longrightarrow} & H_{\mathcal{G}}^p(\eta_\delta) & & &
\end{array}
\tag{8.32}
$$

Now $\nu_\delta \to \eta_\delta$ is a complex vector bundle with a $\mathcal{G}$-action and a circle subgroup that fixes η_δ and acts freely on $\nu_\delta \setminus \eta_\delta$, so by [1, Prop. 13.4], β is injective. It follows that is α is injective as well, and hence the sequence splits. The second statement follows by Remark 8.3.19 and the same argument as above. $\qquad \square$

It remains to prove (8.28) and the remaining cases of (8.26). As noted above, in these cases the negative normal spaces are no longer constant in dimension, and indeed they are not even linear in the fibers. From the point of view of deformation theory, the Kuranishi map near these critical sets is not surjective, and defining an appropriate retraction is more difficult than in the situation just considered. Instead, we resort to the analog of the decomposition used in Section 8.3.3. Let $X''_\delta = X_\delta \setminus \mathrm{pr}^{-1}(\mathcal{A}_{j(\delta)})$. Note that by Lemma 8.2.12, $X''_\delta \subset X'_\delta$. We will prove the following

Proposition 8.3.21 *Suppose* $\delta \in \Delta_{\tau,d}^+ \cap (\tau - d/2, \tau]$. *Then*

$$H_{\mathcal{G}}^*(X_\delta, X''_\delta) \cong H_{\mathcal{G}}^*(\nu_{I,\delta}, \nu''_{I,\delta})) \tag{8.33}$$

$$H_{\mathcal{G}}^*(X'_\delta, X''_\delta) \cong H_{\mathcal{G}}^*(\nu'_{I,\delta}, \nu''_{I,\delta}) \tag{8.34}$$

Proof of (8.33) By [1] and (8.19), it suffices to prove

$$H_{\mathcal{G}}^*(X_\delta, X''_\delta) \cong H_{\mathcal{G}}^*(X_{j(\delta)}^{\mathcal{A}}, X_{j(\delta)-1}^{\mathcal{A}}) \tag{8.35}$$

We first note that the pair (X_δ, X''_δ) is not necessarily invariant under scaling $t\Phi$, $t \to 0$, in particular because of the strata in $\Delta_{\tau,d}^-$ (cf. Lemma 8.2.12). However, if we set

$$\widehat{X}_\delta = X_\delta \cup \bigcup_{\delta' \le \delta,\, \delta' \in \Delta_{\tau,d}^-} X_{\delta'+\tau}^{\mathcal{A}}, \quad \widehat{X}_\delta'' = \widehat{X}_\delta \setminus \mathrm{pr}^{-1}(\mathcal{A}_{j(\delta)})$$

then by excision on the closed subset

$$\bigcup_{\substack{j(\delta)-\tau < \delta' \le \delta \\ \delta' \in \Delta_{\tau,d}^-}} \mathcal{A}_{\delta'+\tau}$$

it follows that $H_{\mathcal{G}}^*(X_\delta, X_\delta'') = H_{\mathcal{G}}^*(\widehat{X}_\delta, \widehat{X}_\delta'')$. Then for the pair $(\widehat{X}_\delta, \widehat{X}_\delta'')$, projection to $\mathcal{A}$ is a deformation retraction (by scaling the section Φ), and we have

$$H_{\mathcal{G}}^*(X_\delta, X_\delta'') = H_{\mathcal{G}}^*(\mathrm{pr}(\widehat{X}_\delta), \mathrm{pr}(\widehat{X}_\delta'')) \tag{8.36}$$

Next, let

$$\mathcal{K}_\delta = \mathrm{pr}\Big(\widehat{X}_{(\tau-d/2)} \cup \bigcup_{\delta' \le \delta,\, \delta' \in \Delta_{\tau,d}^-} \mathcal{B}_{\delta'} \cup \bigcup_{\delta' < \tau-d/2,\, \delta' \in \Delta_{\tau,d}^+} \mathcal{B}_{\delta'}\Big) \cap \bigcup_{k > j(\delta)} \mathcal{A}_k$$

Note that $\mathcal{K}_\delta \subset \mathrm{pr}(\widehat{X}_\delta'')$. We claim that it is actually a closed subset of $\mathrm{pr}(\widehat{X}_\delta)$. Indeed, suppose $(A_i, \Phi_i) \in X_{(\tau-d/2)}$, $(A_i, \Phi_i) \to (A, \Phi) \in \widehat{X}_\delta$, and suppose that $\mu_+(A_i) > j(\delta)$ for each i. By semicontinuity, it follows that $\mu_+(A) > j(\delta)$. On the other had, either $A \in \mathcal{K}_\delta$ or $(A, \Phi) \in \mathcal{B}_{\delta'}$, $\tau - d/2 < \delta' \le \delta$ and $\delta' \in \Delta_{\tau,d}^+$. But by Lemma 8.2.12, this would imply $A \in \mathcal{A}_{j(\delta')}$; which is a contradiction, since $j(\delta') \le j(\delta)$. It follows that the latter cannot occur, and hence, $\mathcal{K}_\delta$ is closed. Similarly,

$$\mathrm{pr}(\widehat{X}_\delta) = \mathrm{pr}\Big(\widehat{X}_{(\tau-d/2)} \cup \bigcup_{\delta' \le \delta,\, \delta' \in \Delta_{\tau,d}^-} \mathcal{B}_{\delta'} \cup \bigcup_{\delta' < \tau-d/2,\, \delta' \in \Delta_{\tau,d}^+} \mathcal{B}_{\delta'}\Big)$$

$$\cup \bigcup_{d/2 < k \le j(\delta)} \mathcal{A}_k \cup \bigcup_{\tau-d/2 < \delta' \le \delta,\, \delta' \in \Delta_{\tau,d}^+} \mathrm{pr}(\mathcal{B}_\delta)$$

$$= \mathcal{K}_\delta \cup \bigcup_{d/2 \le k \le j(\delta)} \mathcal{A}_k$$

and the union is disjoint. It follows also that

$$\mathrm{pr}(\widehat{X}_\delta'') = \mathcal{K}_\delta \cup \bigcup_{d/2 \le k < j(\delta)} \mathcal{A}_k$$

Hence, $\mathrm{pr}(\widehat{X}_\delta) \setminus \mathcal{K}_\delta = X_{j(\delta)}^{\mathcal{A}}$, $\mathrm{pr}(\widehat{X}_{j(\delta)}'') \setminus \mathcal{K}_\delta = X_{j(\delta)-1}^{\mathcal{A}}$, and (8.35) follows from (8.36) by excision. $\qquad\square$

Proof of (8.34) First consider the case $\delta \in \Delta_{\tau,d}^+ \cap (\tau - d/2, 2\tau - d)$. We have

$$X_\delta'' = \left(X_{(\tau-d/2)} \cup \bigcup_{\delta' \leq \delta,\, \delta' \in \Delta_{\tau,d}^-} \mathcal{B}_{\delta'} \cup \bigcup_{\delta' < \tau - d/2,\, \delta' \in \Delta_{\tau,d}^+} \mathcal{B}_{\delta'}\right) \setminus \mathrm{pr}^{-1}(\mathcal{A}_{j(\delta)})$$

$$\cup \bigcup_{d/2 < k < j(\delta)} \mathcal{A}_k \cup \bigcup_{\tau - d/2 < \delta' < \delta,\, \delta' \in \Delta_{\tau,d}^+} \mathcal{B}_\delta$$

whereas $X_\delta' = X_{\delta_1} \cup \mathcal{B}_\delta$, where δ_1 is the predecessor of δ in $\Delta_{\tau,d}$. Also, $X_\delta'' = X_{\delta_1} \setminus \mathrm{pr}^{-1}(\mathcal{A}_{j(\delta)})$. We then have the following diagram

$$\begin{array}{ccccccc}
\cdots \longrightarrow & H_{\mathcal{G}}^p(X_\delta', X_\delta'') & \longrightarrow & H_{\mathcal{G}}^p(X_\delta') & \longrightarrow & H_{\mathcal{G}}^p(X_\delta'') & \longrightarrow \cdots \quad (8.37)\\
& \downarrow f & & \downarrow g & & \downarrow \cong & \\
\cdots \longrightarrow & H_{\mathcal{G}}^p(X_{\delta_1}, X_\delta'') & \longrightarrow & H_{\mathcal{G}}^p(X_{\delta_1}) & \longrightarrow & H_{\mathcal{G}}^p(X_\delta'') & \longrightarrow \cdots
\end{array}$$

where f and g are induced by the inclusion $X_{\delta_1} \hookrightarrow X_\delta'$. By Lemma 8.3.20 and (8.27) (see Remark 8.3.19), it follows that g is surjective and

$$\ker g = H_{\mathcal{G}}^*(\nu_{II,\delta}, \nu_{II,\delta}') \simeq H_{\mathcal{G}}^{*-2j(\delta)}(\mathcal{B}_\delta) \simeq H_{S^1}^{*-2j(\delta)}(S^{j(\delta)} M \times J_{d-j(\delta)} M)$$

by Thom isomorphism. Chasing through the diagram, it follows that f is also surjective with the same kernel. We conclude that

$$H_{\mathcal{G}}^*(X_\delta', X_\delta'') \simeq H_{\mathcal{G}}^*(X_{\delta_1}, X_\delta'') \oplus H_{S^1}^{*-2j(\delta)}(S^{j(\delta)} M \times J_{d-j(\delta)} M) \quad (8.38)$$

It remains to compute the first factor on the right hand side. To begin, notice that

$$\bigcup_{d/2 < k < j(\delta)} \mathcal{A}_k \cup \bigcup_{\tau - d/2 < \delta' < \delta,\, \delta' \in \Delta_{\tau,d}^-} \mathcal{B}_{\delta'} \cup \bigcup_{\tau - d/2 < \delta' < \delta,\, \delta' \in \Delta_{\tau,d}^+} \mathcal{B}_\delta$$

is contained in X_δ'' and closed in X_{δ_1}. It follows by excision that

$$H_{\mathcal{G}}^*(X_{\delta_1}, X_\delta'') \simeq H_{\mathcal{G}}^*(X_{\tau-d/2}, X_{\tau-d/2} \setminus \mathrm{pr}^{-1}(\mathcal{A}_{j(\delta)}))$$

Next, we observe that

$$\mathcal{A}_{ss} \cup \bigcup_{\delta' < \tau - d/2,\, \delta' \in \Delta_{\tau,d}^-} \mathcal{B}_{\delta'}$$

is contained in $X_{\tau-d/2} \setminus \mathrm{pr}^{-1}(\mathcal{A}_{j(\delta)})$ and closed in $X_{\tau-d/2}$. This is clear for $\mathcal{A}_{ss}$. More generally, if (E, Φ) in this set and $\Phi \not\equiv 0$, then $\mu_+(E) > \tau > j(\delta)$, and elements in the strata of type $\mathbf{II}^-$ cannot specialize to points in $\mathbf{II}^+$. Again applying excision, we have

$$H_{\mathcal{G}}^*(X_{\delta_1}, X_\delta'') \simeq H_{\mathcal{G}}^*(Y_\delta, Y_\delta \setminus \mathrm{pr}^{-1}(\mathcal{A}_{j(\delta)}))$$

where

$$Y_\delta = \mathcal{B}_{ss}^\tau \cup \bigcup_{0 < \delta' \leq \tau - d/2,\, \delta' \in \Delta_{\tau,d}^+} \mathcal{B}_{\delta'}$$

We make a third excision of the closed set

$$\bigcup_{\tau - j(\delta) < \delta' \leq \tau - d/2,\, \delta' \in \Delta_{\tau,d}^+} \mathcal{B}_{\delta'}$$

and a final excision of the subset

$$\mathcal{D}_\delta = \left\{ \mathcal{B}_{ss}^\tau \cup \bigcup_{0 < \delta' \leq \tau - j(\delta),\, \delta' \in \Delta_{\tau,d}^+} \mathcal{B}_\delta \right\} \cap \left(\bigcup_{k > j(\delta)} \mathrm{pr}^{-1}(\mathcal{A}_k) \right)$$

Notice that

$$\left\{ \mathcal{B}_{ss}^\tau \cup \bigcup_{0 < \delta' \leq \tau - j(\delta),\, \delta' \in \Delta_{\tau,d}^+} \mathcal{B}_\delta \right\} \setminus \mathcal{D}_\delta = \mathcal{B}_{ss}^{j(\delta)}$$

We conclude that

$$H_\mathcal{G}^*(X_{\delta_1}, X_\delta'') \simeq H_\mathcal{G}^*(\mathcal{B}_{ss}^{j(\delta)}, \mathcal{B}_{ss}^{j(\delta)} \setminus \mathrm{pr}^{-1}(\mathcal{A}_{j(\delta)}))$$

Choose $\varepsilon > 0$ small, and let $\tau' = j(\delta) - \varepsilon$. Then with respect to the τ'-stratification, the right hand side above is $\simeq H_\mathcal{G}^*(\mathcal{B}_{ss}^{\tau'} \cup \mathcal{B}_\varepsilon^{\tau'}, \mathcal{B}_{ss}^{\tau'})$ where $\varepsilon \in \Delta_{\tau'}^-$ is the lowest τ'-critical set. Since $\varepsilon < \tau' - d/2$, it follows from Lemma 8.3.20 that the long exact sequence (8.9) splits for this stratum. Hence, we have

$$H_\mathcal{G}^*(X_{\delta_1}, X_\delta'') \simeq H_\mathcal{G}^*(\mathcal{B}_{ss}^{\tau'} \cup \mathcal{B}_\varepsilon^{\tau'}, \mathcal{B}_{ss}^{\tau'}) \simeq H_{S^1}^{*-2(2j(\delta)-d+g-1)}(S^{d-j(\delta)}M \times J_{j(\delta)}(M)) \tag{8.39}$$

(notice that $j_{\tau'}(\varepsilon) = j_\tau(\delta)$). Eqs. (8.38) and (8.39), combined with Proposition 8.3.17, complete the proof. In case $\delta \notin I_{\tau,d}$, note that by definition $H_\mathcal{G}^*(X_\delta', X_\delta'') \simeq H_\mathcal{G}^*(X_{\delta_1}, X_\delta'')$. The part of the proof following (8.38) now applies verbatim to this case. $\qquad \square$

Proof of Theorem 8.3.18 For $\delta \notin \Delta_{\tau,d}^+ \cap [\tau - d/2, \tau]$, or $\delta = \tau - d/2$ and $d > 4g - 4$, we have proven the result directly (see the discussion following Theorem 8.3.18 and also Remark 8.3.19). For $\delta \in \Delta_{\tau,d}^+ \cap (\tau - d/2, \tau]$, the result follows from Proposition 8.3.21 and the five lemma. $\qquad \square$

8.3.6 Perfection of the stratification for large degree

Note that Lemma 8.3.20 shows that the long exact sequence (8.9) splits for all $\delta \notin \Delta_{\tau,d}^+ \cap [\tau - d/2, \tau]$, and also for $\delta = \tau - d/2$ if $d > 4g - 4$. Therefore it remains to show that (8.9) splits for $\delta \in \Delta_{\tau,d}^+ \cap (\tau - d/2, \tau]$.

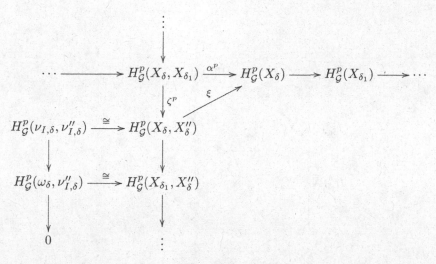

Figure 8.2

Firstly we consider the case where $\delta \in \Delta_{\tau,d}^+ \cap [2\tau - d, \tau]$, which corresponds to a stratum of type $\mathbf{I}_b$. Proposition 8.3.21 shows that the vertical long exact sequence splits and the map ξ is injective in the following commutative diagram (Figure 8.2).

Therefore the map α^p is injective, and so the horizontal long exact sequence splits also.

Next, suppose $\delta \in \Delta_{\tau,d}^+ \cap (\tau - d/2, 2\tau - d)$. For this we need the following lemma.

Lemma 8.3.22 *When* $\delta \in \Delta_{\tau,d}^+ \cap (\tau - d/2, 2\tau - d)$, *then the isomorphisms* $H_{\mathcal{G}}^*(X_\delta, X_\delta'') \cong H_{\mathcal{G}}^*(\nu_{I,\delta}, \nu_{I,\delta}'')$ *and* $H_{\mathcal{G}}^*(X_{\delta_1}, X_\delta'') \cong H_{\mathcal{G}}^*(\omega_\delta, \nu_{I,\delta}'')$ *in equivariant cohomology are induced by an inclusion of triples* $(\nu_{I,\delta}, \omega_\delta, \nu_{I,\delta}'') \hookrightarrow (X_\delta, X_{\delta_1}, X_\delta'')$.

Proof The first isomorphism is contained in (8.33). To see the second isomorphism, note that the results of the last section show that $H_{\mathcal{G}}^*(X_{\delta_1}, X_\delta'') \cong H_{\mathcal{G}}^*(\mathcal{B}_{ss}^{\tau'} \cup \mathcal{B}_\varepsilon^{\tau'}, \mathcal{B}_{ss}^{\tau'})$, where $\varepsilon \in \Delta_{\tau'}^-$ is the lowest τ' critical set. Excise all but a neighborhood of $\mathcal{B}_\varepsilon^{\tau'}$, and deformation retract Φ so that $\|\Phi\|$ is small. Call these new sets W and W_0, respectively. Then

$$H_{\mathcal{G}}^*(\mathcal{B}_{ss}^{\tau'} \cup \mathcal{B}_\varepsilon^{\tau'}, \mathcal{B}_{ss}^{\tau'}) \cong H_{\mathcal{G}}^*(W, W_0)$$

Since $\Phi \neq 0$, then we can apply Lemma 8.2.16 to the slices within the spaces W and W_0, and the resulting spaces are homeomorphic to ω_δ and $\nu_{I,\delta}''$ respectively. $\square$

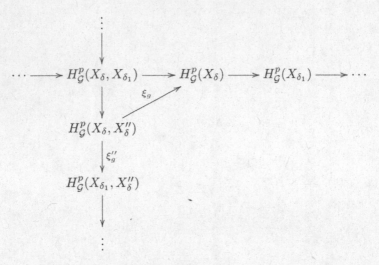

Figure 8.3

The previous lemma together with the surjection $\xi'' : H_{\mathcal{G}}^*(\nu_{I,\delta}^-, \nu_{I,\delta}'') \to H_{\mathcal{G}}^*(\omega_\delta, \nu_{I,\delta}'')$ from (8.22) implies that the map ξ_g'' is surjective in the following commutative diagram (Figure 8.3).

The isomorphism (8.35) together with the results of [1] show that the map ξ_g is injective, and so the same argument as before shows that the horizontal long exact sequence splits.

8.3.7 The case of low degree

By the results of the previous section, there is only one critical stratum unaccounted for on the way to completing the proof of Theorem 8.3.11 for $1 \le d \le 4g - 4$. Namely, we need to analyze what happens when we attach the minimal Yang-Mills stratum $\mathcal{A}_{ss}$, which is the lowest critical set of Type **I**. More precisely, from (8.4), we need to show that the inclusion $X_{\tau-d/2}' \hookrightarrow X_{\tau-d/2}$ induces a surjection in $\mathcal{G}$-equivariant rational cohomology for all $\tau \in (d/2, d)$. Notice that by (8.5), $X_{\tau-d/2}' = X_{\delta_1}$ for d odd, so this is precisely what we need to prove; and if d is even, then the above statement together with Lemma 8.3.20 will prove that $X_{\delta_1} \hookrightarrow X_{\tau-d/2}$ induces a surjection in $\mathcal{G}$-equivariant rational cohomology in this case as well.

In low degree, the negative normal directions exist only over a Brill-Noether subset of $\mathcal{A}_{ss}$, whose cohomology is unknown, and the dimension of the fiber jumps in a complicated way; it is not even clear that there is a good Morse-Bott lemma of the type (8.28) in this case.

Hence, in order to prove surjectivity in this case we will use an indirect argument via embeddings of the space of pairs of degree d into corresponding pairs of larger degree. More precisely, this is defined as follows. Choose a point $p \in M$, and let $\mathcal{O}(p)$ denote the holomorphic line bundle with divisor p. We also choose a hermitian metric on $\mathcal{O}(p)$. Choose a holomorphic section σ_p of $\mathcal{O}(p)$ with a simple zero at p. Note that σ_p is unique up to a nonzero multiple. A holomorphic (and hermitian) structure on the complex vector bundle E induces one on the bundle $\widetilde{E} = E \otimes \mathcal{O}(p)$. Moreover, if $\Phi \in H^0(E)$, then $\widetilde{\Phi} = \Phi \otimes \sigma_p \in H^0(\widetilde{E})$. The unitary gauge group $\mathcal{G}$ of E is canonically isomorphic to that of $\widetilde{E}$. Hence, we have a $\mathcal{G}$-equivariant embedding $\mathcal{B}(E) \hookrightarrow \mathcal{B}(\widetilde{E})$. For simplicity, we will use the notation $\mathcal{B} = \mathcal{B}(E)$ and $\widetilde{\mathcal{B}} = \mathcal{B}(\widetilde{E})$.

Let $\tilde{d} = d + 2$ and $\tilde{\tau} = \tau + 1$. Then we note the following properties:

$$\deg \widetilde{E} = \tilde{d} \quad \Delta_{\tilde{\tau}, \tilde{d}} = \Delta_{\tau, d}$$

$$\deg \widetilde{\Phi} = \deg \Phi + 1 \quad I_{\tilde{\tau}, \tilde{d}} = I_{\tau, d}$$

$$\mu_+(\widetilde{E}) = \mu_+(E) + 1 \quad j_{\tilde{\tau}, \tilde{d}}(\delta) = j_{\tau, d} + 1$$

It follows easily that the inclusion respects the Harder-Narasimhan stratification, i.e. for all $\delta \in \Delta_{\tau, d}, \mathcal{B}_\delta \hookrightarrow \widetilde{\mathcal{B}}_\delta, X_\delta \hookrightarrow \widetilde{X}_\delta$, and $X_\delta' \hookrightarrow \widetilde{X}_\delta'$, where the tilde's have the obvious meaning. In particular, if we fix $\tau_{max} = d - \varepsilon$, for ε small, then $\mathcal{B}_{ss}^{\tau_{max}} \hookrightarrow \widetilde{\mathcal{B}}_{ss}^{\tau_{max}}$. Notice that while $\mathcal{B}_{ss}^{\tau_{max}}$ gives the "last" moduli space in the sense that there are no critical values between τ_{max} and d (provided ε is sufficiently small), $\widetilde{\mathcal{B}}_{ss}^{\tilde{\tau}_{max}}$ gives the "second to last" moduli space in the sense that there is precisely one critical value between $\tilde{\tau}_{max}$ and $\tilde{d}$.

Lemma 8.3.23 *The inclusion* $\mathcal{B}_{ss}^{\tau_{max}} \hookrightarrow \widetilde{\mathcal{B}}_{ss}^{\tilde{\tau}_{max}}$ *induces a surjection in* $\mathcal{G}$-*equivariant rational cohomology.*

Proof Since τ is generic, it suffices to prove the result on the level of moduli spaces, i.e. that the inclusion $\iota : \mathfrak{M}_{\tau_{max}, d} \hookrightarrow \widetilde{\mathfrak{M}}_{\tilde{\tau}_{max}, \tilde{d}}$ induces a surjection in cohomology. Consider the determinant map $(E, \Phi) \mapsto \det E$. We have the following diagram

$$
\begin{array}{ccc}
\mathfrak{M}_{\tau_{max}, d} & \stackrel{\iota}{\longrightarrow} & \widetilde{\mathfrak{M}}_{\tilde{\tau}_{max}, \tilde{d}} \\
\downarrow{\scriptstyle \det} & & \downarrow{\scriptstyle \det} \\
J_d(M) & \stackrel{J}{\longrightarrow} & J_{\tilde{d}}(M)
\end{array}
\tag{8.40}
$$

Now $\mathfrak{M}_{\tau_{max}, d}$ is the projectivization of a vector bundle (cf. [20]). Hence, by the Leray–Hirsch theorem its cohomology ring is generated by the embedding $(\det)^*(H^*(J_d(M)))$, and a 2-dimensional class generating the cohomology of the fiber. Since $\iota^*(\det)^* = (\det)^* J^*$, and J^* is an isomorphism, it follows that ι^*

is surjective onto $(\det)^*(H^*(J_d(M)))$. It remains to show that the 2-dimensional class is in the image of ι^*. But since ι is holomorphic and $\mathfrak{M}_{\tilde\tau_{max},\tilde{d}}$ is projective, the Kähler class of $\mathfrak{M}_{\tilde\tau_{max},\tilde{d}}$ restricted to the image generates the cohomology of the fiber.   □

Lemma 8.3.24 *Suppose* $\delta \in \Delta_{\tau_{max},d}$, $\delta < \tau_{max} - d/2$. *Then the inclusion* $X_\delta \hookrightarrow \widetilde{X}_\delta$ *induces a surjection in* $\mathcal{G}$-*equivariant rational cohomology. The same holds for* $X'_{\tau-d/2} \hookrightarrow \widetilde{X}'_{\tilde\tau-\tilde{d}/2}$.

Proof By Lemma 8.3.23, the result holds for the semistable stratum. Fix $\delta < \tau - d/2$, and let δ_1 be its predecessor in $\Delta_{\tau_{max},d}$. By induction, we may assume the result holds for δ_1. By Lemma 8.3.20 we have the following diagram:

$$
\begin{array}{ccccccccc}
0 & \longrightarrow & H_\mathcal{G}^p(\widetilde{X}_\delta, \widetilde{X}_{\delta_1}) & \longrightarrow & H_\mathcal{G}^p(\widetilde{X}_\delta) & \longrightarrow & H_\mathcal{G}^p(\widetilde{X}_{\delta_1}) & \longrightarrow & 0 \\
& & \downarrow{\scriptstyle f} & & \downarrow{\scriptstyle g} & & \downarrow{\scriptstyle h} & & \\
0 & \longrightarrow & H_\mathcal{G}^p(X_\delta, X_{\delta_1}) & \longrightarrow & H_\mathcal{G}^p(X_\delta) & \longrightarrow & H_\mathcal{G}^p(X_{\delta_1}) & \longrightarrow & 0
\end{array}
\tag{8.41}
$$

By the inductive hypothesis, h is surjective. On the other hand, by (8.26) and (8.29), surjectivity of f is equivalent to surjectivity of the map $H_\mathcal{G}^*(\tilde\eta_\delta) \to H_\mathcal{G}^*(\eta_\delta)$. From the description of critical sets (cf. Proposition 8.3.2), this map is induced by the inclusion $S^{j(\delta)}M \hookrightarrow S^{j(\delta)+1}M$. Surjectivity then follows by the argument in [8, Sect. 4]. Since both f and h are surjective, so is g. The result for any $\delta < \tau - d/2$ now follows by induction. If d is even, the exact same argument, with $\delta_1 = $ the predecessor of $\tau - d/2$, proves the statement for $X'_{\tau-d/2}$ as well.   □

Lemma 8.3.25 *Suppose the inclusion* $\widetilde{X}'_{\tilde\tau_{max}-\tilde{d}/2} \hookrightarrow \widetilde{X}_{\tilde\tau_{max}-\tilde{d}/2}$ *induces a surjection in* $\mathcal{G}$-*equivariant rational cohomology. Then the same is true for the inclusion* $X'_{\tau_{max}-d/2} \hookrightarrow X_{\tau_{max}-d/2}$.

Proof Consider the diagram

$$
\begin{array}{ccccc}
H_\mathcal{G}^p(\widetilde{X}_{\tilde\tau_{max}-\tilde{d}/2}) & \longrightarrow & H_\mathcal{G}^p(\widetilde{X}'_{\tilde\tau_{max}-\tilde{d}/2}) & \longrightarrow & 0 \\
\downarrow & & \downarrow{\scriptstyle h} & & \\
H_\mathcal{G}^p(X_{\tau_{max}-d/2}) & \longrightarrow & H_\mathcal{G}^p(X'_{\tau_{max}-d/2}) & \longrightarrow & \cdots
\end{array}
\tag{8.42}
$$

By Lemma 8.3.24, h is surjective. The result then follows immediately.   □

Lemma 8.3.26 *Suppose the inclusion* $X'_{\tau-d/2} \hookrightarrow X_{\tau-d/2}$ *induces a surjection in* $\mathcal{G}$-*equivariant rational cohomology for* $\tau = \tau_{max}$. *Then the same is true for* *all* $\tau \in (d/2, d)$. *Moreover,* dim $H_{\mathcal{G}}^p(X_{\tau-d/2}, X'_{\tau-d/2})$ *is independent of* τ *for all* p.

Proof The sets $X'_{\tau-d/2}, X_{\tau-d/2}$ remain unchanged for τ in a connected component of $(d/2, d) \setminus C_d$, where C_d is given in (8.6). Fix $\tau_c \in C_d, 2\tau_c - d/2 = k \in \mathbb{Z}$, and let $\tau_l < \tau_c < \tau_r$ be in components $(d/2, d) \setminus C_d$ containing τ_c in their closures. Let $\delta^{l,r} = 2\tau_c - d/2 - \tau_{l,r}$. Note that $\delta^{l,r} \in \Delta_{\tau_{l,r},d}^-$, $\delta^l > \tau_l - d/2$, and $\delta^r < \tau_r - d/2$. Also, we claim

$$X_{\tau_r-d/2} = X_{\tau_l-d/2} \cup \mathcal{B}_{\delta^l}^{\tau_l} \ , \ X'_{\tau_r-d/2} = X'_{\tau_l-d/2} \cup \mathcal{B}_{\delta^l}^{\tau_l} \qquad (8.43)$$

To see this, we refer to Figure 8.1 and the discussion preceding it. Under the map $\Delta_{\tau_l,d} \to \Delta_{\tau_r,d}$, $\delta^l \mapsto \delta^r$ and $\tau_l - d/2 \mapsto \tau_r - d/2$. The claim then follows if we show that δ^r is the predecessor of $\tau_r - d/2$ in $\Delta_{\tau_r,d}$, and δ^l is the successor of $\tau_l - d/2$ in $\Delta_{\tau_l,d}$ (see Figure 8.1). So suppose $\delta \in \Delta_{\tau_r,d}, \delta < \tau_r - d/2$. By Remark 8.2.7, we may assume $\delta \in \Delta_{\tau_r,d}^-$. Write $\delta + \tau_r = \ell \in \mathbb{Z}$. Then $\ell \le 2\tau_r - d/2$, which implies $\ell \le k$, and $\delta \le \delta_r$. The reasoning is similar for δ_l.

Now since the result holds by assumption for τ_{max}, we may assume by induction that the result holds for $\tau \ge \tau_r$. Then we have

$$0 \longrightarrow H_{\mathcal{G}}^p(X_{\tau_r-d/2}, X'_{\tau_r-d/2}) \longrightarrow H_{\mathcal{G}}^p(X_{\tau_r-d/2}) \longrightarrow H_{\mathcal{G}}^p(X'_{\tau_r-d/2}) \longrightarrow 0$$

$$\Big\downarrow f \qquad\qquad\qquad \Big\downarrow g \qquad\qquad\qquad \Big\downarrow h$$

$$\cdots \longrightarrow H_{\mathcal{G}}^p(X_{\tau_l-d/2}, X'_{\tau_l-d/2}) \longrightarrow H_{\mathcal{G}}^p(X_{\tau_l-d/2}) \longrightarrow H_{\mathcal{G}}^p(X'_{\tau_l-d/2}) \longrightarrow \cdots$$

$$(8.44)$$

By (8.43) and the proof of Lemma 8.3.24, h is surjective. Hence, the lower long exact sequence must split. Moreover, g is surjective as well, and ker $g = $ ker h. As a consequence, f must be an isomorphism. The result now follows by induction. $\qquad\square$

Proof of Theorems 8.3.11 and 8.1.2 We proceed by induction as follows. First, if $d > 4g - 4$, then by Lemma 8.3.20, the hypothesis of Lemma 8.3.25 is satisfied. It then follows from Lemma 8.3.26 that the inclusion $X'_{\tau-d/2} \hookrightarrow X_{\tau-d/2}$ induces a surjection in $\mathcal{G}$-equivariant rational cohomology for any τ. In particular, this is true for the value $\tilde{\tau}_{max}$ corresponding to degree $d - 2$. Hence, the inductive hypothesis holds, and the result is proven for all d. Kirwan surjectivity follows immediately. $\qquad\square$

Proof of Theorem 8.1.3 This follows from Kirwan surjectivity, but more generally we prove this on each stratum. Clearly it suffices to prove the result for

$k = 1$. Since the gauge groups for E and $\widetilde{E}$ are canonically isomorphic, it suffices by induction to show that if the result holds for the inclusion $X_\delta \hookrightarrow \widetilde{X}_\delta$, then it also holds for $X_{\delta_1} \hookrightarrow \widetilde{X}_{\delta_1}$, where δ_1 is the predecessor of δ in $\Delta_{\tau,d}$. By Theorem 8.3.11, the diagram (8.41) holds for *all* δ. It follows that if g is surjective, then so is h. This completes the proof. $\square$

8.4 Cohomology of moduli spaces

8.4.1 Equivariant cohomology of τ-semistable pairs

The purpose of this section is to complete the calculation of the $\mathcal{G}$-equivariant Poincaré polynomial of $\mathcal{B}_{ss}^\tau$. First we consider the case where τ is generic. Choose an integer N, $d/2 < N \leq d$, and let $\tau \in (\max\{d/2, N-1\}, N)$. Then the different allowable values of δ for each type of stratum and the cohomology are as follows (see Proposition 8.3.2).

($\mathbf{I}_a$) There is one stratum $\mathbf{I}_a^{d/2}$ corresponding to $\mathcal{A}_{ss}$ (indexed by $j = d/2$), and by Lemma 8.3.26 the contribution $\mathbf{I}_a^{d/2}(t)$ to the Poincaré polynomial is independent of τ. For $d > 4g - 4$ it follows from (8.30) that

$$\mathbf{I}_a^{d/2}(t) = \frac{t^{2d+4-4g}}{(1-t^2)} P_t^{\overline{\mathcal{G}}}(\mathcal{A}_{ss}) \tag{8.1}$$

where $\overline{\mathcal{G}}$ is defined in [1, p. 577], We compute $\mathbf{I}_a^{d/2}(t)$ in general in Lemma 8.4.3 below. The remaining strata are indexed by integers $j = j(\delta) = \mu_+$ such that $d/2 < j \leq N-1$ and $\delta = j - d + \tau$. The contribution to the $\mathcal{G}$-equivariant Poincaré polynomial is

$$\begin{aligned}
\mathbf{I}_a^j(t) &= \frac{t^{2(2j(\delta)-d+g-1)}}{(1-t^2)^2} P_t \left(J_{j(\delta)}(M) \times J_{d-j(\delta)}(M) \right) \\
&\quad - \frac{t^{2j(\delta)}}{1-t^2} P_t \left(S^{j(\delta)} M \times J_{d-j(\delta)}(M) \right) \\
&\quad - \frac{t^{2(2j(\delta)-d+g-1)}}{1-t^2} P_t \left(S^{d-j(\delta)} M \times J_{j(\delta)}(M) \right)
\end{aligned} \tag{8.2}$$

($\mathbf{I}_b$) There are an infinite number of strata indexed by integers $j = j(\delta) = \mu_+$ such that $N \leq j$ and $\delta = j - d + \tau$. The contribution is

$$\begin{aligned}
\mathbf{I}_b^j(t) &= \frac{t^{2(2j(\delta)-d+g-1)}}{(1-t^2)^2} P_t \left(J_{j(\delta)}(M) \times J_{d-j(\delta)}(M) \right) \\
&\quad - \frac{t^{2(2j(\delta)-d+g-1)}}{1-t^2} P_t \left(S^{d-j(\delta)} M \times J_{j(\delta)}(M) \right)
\end{aligned} \tag{8.3}$$

(II^+) These strata are indexed by integers $j = j(\delta) = \deg \Phi = \deg L_1$ such that $d - N + 1 \le j \le N - 1$, and $\delta = j - d + \tau$. The contribution is

$$\text{II}_j^+(t) = \frac{t^{2j(\delta)}}{1 - t^2} P_t \left(S^{j(\delta)} M \times J_{d-j(\delta)}(M) \right) \qquad (8.4)$$

(II^-) These strata are indexed by integers $j = d - j(\delta)$ such that $0 \le j \le d - N$, where $\delta = j(\delta) - \tau = d - j - \tau$, and the contribution is

$$\text{II}_j^-(t) = \frac{t^{2(2j(\delta)-d+g-1)}}{1 - t^2} P_t \left(S^{d-j(\delta)} M \times J_{j(\delta)}(M) \right) \qquad (8.5)$$

Then we have

Theorem 8.4.1 *For $\tau \in (\max\{d/2, N - 1\}, N)$,*

$$P_t(\mathfrak{M}_{\tau,d}) = P_t^{\mathcal{G}}(\mathcal{B}_{ss}^{\tau})$$

$$= P_t(B\mathcal{G}) - \mathbf{I}_a^{d/2}(t) - \sum_{j=\lfloor d/2+1 \rfloor}^{N-1} \mathbf{I}_a^j(t) - \sum_{j=N}^{\infty} \mathbf{I}_b^j(t)$$

$$- \sum_{j=0}^{d-N} \text{II}_j^-(t) - \sum_{j=d-N+1}^{N-1} \text{II}_j^+(t)$$

Proof By Theorem 8.3.11 we have

$$P_t^{\mathcal{G}}(\mathcal{B}_{ss}^{\tau}) = P_t(B\mathcal{G}) - \sum_{\delta \in \Delta_{\tau,d} \setminus \{0\}} P_t^{\mathcal{G}}(X_{\delta}, X_{\delta_1})$$

If $\delta \notin \Delta_{\tau,d}^+ \cap (\tau - d/2, \tau]$, then by the Morse–Bott lemma (8.26) and (8.29),

$$P_t^{\mathcal{G}}(X_{\delta}, X_{\delta_1}) = \frac{t^{2\sigma(\delta)}}{1 - t^2} P_t^{\mathcal{G}}(\eta_{\delta})$$

where $\sigma(\delta)$ is given in Corollary 8.3.14. If $\delta \in \Delta_{\tau,d}^+ \cap (\tau - d/2, \tau]$ then by Section 8.3.6,

$$P_t^{\mathcal{G}}(X_{\delta}, X_{\delta_1}) = P_t^{\mathcal{G}}(X_{\delta}, X_{\delta}'') - P_t^{\mathcal{G}}(X_{\delta_1}, X_{\delta}'')$$

The first term on the right hand side is given by (8.35). For the second term, we have

$$H_{\mathcal{G}}^*(X_{\delta_1}, X_{\delta}'') = \begin{cases} H_{\mathcal{G}}^*(\nu_{I,\delta}', \nu_{I,\delta}'') & \delta \in \Delta_{\tau,d}^+ \cap [2\tau - d, \tau] \\ H_{\mathcal{G}}^*(\omega_{\delta}, \nu_{I,\delta}'') & \delta \in \Delta_{\tau,d}^+ \cap (\tau - d/2, 2\tau - d) \end{cases}$$

and the latter cohomology groups have been computed in (8.20) and (8.25). This completes the computation.   □

When the parameter τ is non-generic (i.e. $\tau = N$ for some integer $N \in [d/2, d]$) then the same analysis as above applies, however now there are split solutions to the vortex equations. These correspond to one of the critical sets of type $\mathbf{II}$, where $E = L_1 \oplus L_2$ with $\phi \in H^0(L_1) \setminus \{0\}$, and $\deg L_2 = \tau$. Therefore, the only difference the generic and non-generic case is that we do not count any contribution from the critical set of type $\mathbf{II}^-$ with $j = d - N$. Therefore the Poincaré polynomial is

Theorem 8.4.2 *For* $\tau = N$,

$$P_t^{\mathcal{G}}(\mathcal{B}_{ss}^N) = P_t(B\mathcal{G}) - \mathbf{I}_a^{d/2}(t) - \sum_{j=\lfloor d/2+1 \rfloor}^{N-1} \mathbf{I}_a^j(t) - \sum_{j=N}^{\infty} \mathbf{I}_b^j(t)$$

$$- \sum_{j=0}^{d-N-1} \mathbf{II}_j^-(t) - \sum_{j=d-N+1}^{N-1} \mathbf{II}_j^+(t)$$

Finally, using Theorem 8.4.1, we can give a computation of the remaining term which is as yet undetermined in low degree.

Lemma 8.4.3 *For all* $d \geq 2$,

$$\mathbf{I}_a^{d/2}(t) = \frac{1}{1-t^2} P_t^{\overline{\mathcal{G}}}(\mathcal{A}_{ss}) - \sum_{j=0}^{\lfloor d/2 \rfloor} \frac{t^{2j} - t^{2(d+g-1-2j)}}{1-t^2} P_t(S^j M \times J_{d-j}(M))$$

$$- \begin{cases} 0 & \text{if } d \text{ odd} \\ \dfrac{t^{2g-2}}{(1-t^2)} P_t(S^{d/2} M \times J_{d/2}(M)) & \text{if } d \text{ even} \end{cases}$$

Remark 8.4.4 It can be verified directly that for $d > 4g - 4$, the expression above agrees with (8.1). See the argument of Zagier in [20, pp. 336–7].

Proof of Lemma 8.4.3 Take the special case $N = d$. Then $\mathfrak{M}_{\tau,d}$ is a projective bundle over $J_d(M)$, and so

$$P_t(\mathfrak{M}_{\tau,d}) = \frac{1 - t^{2(d+g-1)}}{1-t^2} P_t(J_d(M))$$

On the other hand, from Theorem 3.1 we have

$$P_t(\mathfrak{M}_{\tau,d}) = P_t(B\mathcal{G}) - \mathbf{I}_a^{d/2}(t) - \sum_{j=\lfloor d/2+1 \rfloor}^{d-1} \mathbf{I}_a^j(t) - \sum_{j=d}^{\infty} \mathbf{I}_b^j(t) - \mathbf{II}_0^-(t) - \sum_{j=1}^{d-1} \mathbf{II}_j^+(t)$$

Now notice that the term $\mathbf{II}_0^-(t)$ is cancelled by the second term in $\mathbf{I}_b^d$. We combine the remaining terms in the sum of $\mathbf{I}_b^j$ with the sum of $\mathbf{I}_a^j$. We have

$$P_t(\mathfrak{M}_{\tau,d}) = P_t(B\mathcal{G}) - \mathbf{I}_a^{d/2}(t) - \sum_{j=\lfloor d/2+1 \rfloor}^{\infty} \frac{t^{2(2j-d+g-1)}}{(1-t^2)^2} P_t(J_j(M) \times J_{d-j}(M))$$

$$+ \sum_{j=\lfloor d/2+1 \rfloor}^{d-1} \frac{t^{2j}}{(1-t^2)} P_t(S^j M \times J_{d-j}(M))$$

$$+ \sum_{j=\lfloor d/2+1 \rfloor}^{d-1} \frac{t^{2(2j-d+g-1)}}{(1-t^2)} P_t(S^{d-j} M \times J_j(M))$$

$$- \sum_{j=1}^{d-1} \frac{t^{2j}}{(1-t^2)} P_t(S^j M \times J_{d-j}(M))$$

$$= P_t(B\mathcal{G}) - \mathbf{I}_a^{d/2}(t) - \sum_{j=\lfloor d/2+1 \rfloor}^{\infty} \frac{t^{2(2j-d+g-1)}}{(1-t^2)^2} P_t(J_j(M) \times J_{d-j}(M))$$

$$+ \sum_{j=\lfloor d/2+1 \rfloor}^{d-1} \frac{t^{2(2j-d+g-1)}}{(1-t^2)} P_t(S^{d-j} M \times J_j(M))$$

$$- \sum_{j=1}^{\lfloor d/2 \rfloor} \frac{t^{2j}}{(1-t^2)} P_t(S^j M \times J_{d-j}(M))$$

Now make the substitution $j \mapsto d - j$ in the second to the last sum, using

$$d - \lfloor d/2 + 1 \rfloor = \begin{cases} d/2 - 1 = \lfloor d/2 \rfloor - 1 & \text{if } d \text{ even} \\ d/2 - 1/2 = \lfloor d/2 \rfloor & \text{if } d \text{ odd} \end{cases}$$

The result now follows from this, [1, Thm. 7.14], and the fact that $P_t(\mathfrak{M}_{\tau,d})$ is equal to the $j = 0$ term in the sum. $\qquad\square$

8.4.2 Comparison with the results of Thaddeus

In [20], Thaddeus computed the Poincaré polynomial of the moduli space using different methods to those of this paper. The idea is to show that when the parameter τ passes a critical value, then the moduli space $\mathfrak{M}_{\tau,d}$ undergoes a birational transformation consisting of a blow-down along a submanifold and a blow-up along a different submanifold (these transformations are known as "flips"). By computing the change in Poincaré polynomial caused by the flips as the parameter crosses the critical values, and also observing that the moduli space is a projective space for one extreme value of τ, Thaddeus computed the Poincaré polynomial of the moduli space for any value of the parameter. In this section we recover this result from Theorem 8.4.1. In the Morse theory picture we see that the critical point structure changes: As τ increases past a critical value then a new critical set appears, and the index may change at existing critical points.

Theorem 8.4.5 *Let* $N \in \mathbb{Z}$, $d/2 < N \le d - 1$. *Then for* $\tau \in (\max(d/2, N-1), N)$,

$$P_t(\mathfrak{M}_{\tau+1,d}) - P_t(\mathfrak{M}_{\tau,d}) = \frac{t^{4N-2d+2g-2} - t^{2d-2N}}{1 - t^2} P_t \left(S^{d-N} M \times J_N(M) \right)$$

(8.6)

As a consequence, the Poincaré polynomial of the moduli space has the form

$$P_t(\mathfrak{M}_{\tau,d}) = \frac{(1+t)^{2g}}{1 - t^2} \operatorname{Coeff}_{x^N} \left(\frac{t^{2d+2g-2-4N}}{xt^4 - 1} - \frac{t^{2N+2}}{x - t^2} \right) \left(\frac{(1 + xt)^{2g}}{(1 - x)(1 - xt^2)} \right)$$

(8.7)

Remark 8.4.6 Let $\mathfrak{M}_{\tau,d}^0$ denote the moduli space where the bundle has fixed determinant (see [20]). The analysis in this paper applies in this case as well. In particular, one obtains

$$P_t(\mathfrak{M}_{\tau+1,d}^0) - P_t(\mathfrak{M}_{\tau,d}^0) = \frac{t^{4N-2d+2g-2} - t^{2d-2N}}{1 - t^2} P_t \left(S^{d-N} M \right) \qquad (8.8)$$

This exactly corresponds to Thaddeus' results for $P_t(\mathbb{P}W_j^+) - P_t(\mathbb{P}W_j^-)$ [20, p. 336], where $j = d - N$.

Proof of Theorem 8.4.5 By Theorem 8.4.1,

$$P_t(\mathfrak{M}_{\tau+1,d}) - P_t(\mathfrak{M}_{\tau,d}) = -\mathbf{I}_a^N(t) + \mathbf{I}_b^N(t) + \mathbf{II}_{d-N}^-(t) - \mathbf{II}_{d-N}^+(t) - \mathbf{II}_N^+(t)$$

(8.9)

Substituting in the results of (8.2), (8.3), (8.5), and (8.4) gives

$$P_t(\mathfrak{M}_{\tau+1,d}) - P_t(\mathfrak{M}_{\tau,d}) = -\frac{t^{2(2N-d+g-1)}}{(1-t^2)^2} P_t\left(J_N(M) \times J_{d-N}(M)\right)$$

$$+ \frac{t^{2N}}{1-t^2} P_t\left(S^N M \times J_{d-N}(M)\right)$$

$$+ \frac{t^{2(2N-d+g-1)}}{1-t^2} P_t\left(S^{d-N} M \times J_N(M)\right)$$

$$+ \frac{t^{2(2N-d+g-1)}}{(1-t^2)^2} P_t\left(J_N(M) \times J_{d-N}(M)\right)$$

$$- \frac{t^{2(2N-d+g-1)}}{1-t^2} P_t\left(S^{d-N} M \times J_N(M)\right)$$

$$+ \frac{t^{2(d-2(d-N)+g-1)}}{1-t^2} P_t\left(S^{d-N} M \times J_N(M)\right)$$

$$- \frac{t^{2(d-N)}}{1-t^2} P_t\left(S^{d-N} M \times J_N(M)\right)$$

$$- \frac{t^{2N}}{1-t^2} P_t\left(S^N M \times J_{d-N}(M)\right)$$

$$= \frac{1}{1-t^2} P_t\left(S^{d-N} M \times J_N(M)\right)\left(t^{4N-2d+2g-2} - t^{2d-2N}\right)$$

as required. Using the results of [16] on the cohomology of the symmetric product, and the fact that $P_t(J_N(M)) = (1+t)^{2g}$, we see that the same method as for the proof of [20, (4.1)] gives equation (8.7). $\qquad\square$

Remark 8.4.7 For τ as above, Theorem 8.4.2 shows that the difference

$$P_t^{\mathcal{G}}(\mathcal{B}_{ss}^N) - P_t(\mathfrak{M}_{\tau,d}) = \mathrm{II}_{d-N}^-(t) = \frac{t^{4N-2d+2g-2}}{1-t^2} P_t\left(S^{d-N} M \times J_N(M)\right)$$

comes from only one critical set; the type II critical set corresponding to a solution of the vortex equations when $\tau = N$. The rest of the terms in (8.9), corresponding to the difference

$$P_t(\mathfrak{M}_{\tau+1,d}) - P_t^{\mathcal{G}}(\mathcal{B}_{ss}^N) = -\mathrm{I}_a^N(t) + \mathrm{I}_b^N(t) - \mathrm{II}_{d-N}^+(t) - \mathrm{II}_N^+(t)$$

$$= -\frac{t^{2N}}{1-t^2} P_t\left(S^N M \times J_{d-N}(M)\right)$$

come from a number of changes that occur in the structure of the critical sets as τ increases past N: the term $-\mathrm{II}_{d-N}^+(t)$ corresponds to the type II critical point that no longer is a solution to the vortex equations, the term $-\mathrm{II}_N^+(t)$ corresponds to the new critical point of type II^+ that appears, and the term $-\mathrm{I}_a^N(t) + \mathrm{I}_b^N(t)$ corresponds to the critical point that changes type from I_b to I_a.

Therefore we see that the changes in the critical set structure as τ crosses the critical value N are localized to two regions of $\mathcal{B}$. The first corresponds to interchanging critical sets of type II^- and type II^+. This is the phenomenon illustrated in Figure 1. The second corresponds to critical sets of type I_a and II^+ that merge to form a single component of type I_b. The terms from the first change exactly correspond to those in (8.6), i.e.

$$\mathrm{II}_{d-N}^-(t) - \mathrm{II}_{d-N}^+(t) = \frac{t^{4N-2d+2g-2} - t^{2d-2N}}{1-t^2} \cdot P_t\left(S^{d-N}M \times J_N(M)\right)$$
$$= P_t(\mathfrak{M}_{\tau+1,d}) - P_t(\mathfrak{M}_{\tau,d}) \ .$$

and the terms from the second change cancel each other, i.e. $\mathrm{I}_b^N(t) - \mathrm{I}_a^N(t) - \mathrm{II}_N^+(t) = 0$.

References

[1] M. F. Atiyah and R. Bott. The Yang-Mills equations over Riemann surfaces. Philos. Trans. Roy. Soc. London Ser. A, 308 (1983), 523–615.

[2] R. Bott. Nondegenerate critical manifolds. Ann. of Math., 60 (1954), no. 2, 248–261.

[3] S. Bradlow. Special metrics and stability for holomorphic bundles with global sections. J. Differential Geom. 33 (1991), no. 1, 169–213.

[4] S. Bradlow and G.D. Daskalopoulos. Moduli of stable pairs for holomorphic bundles over Riemann surfaces. Internat. J. Math. 2 (1991), no. 5, 477–513.

[5] S. Bradlow, G.D. Daskalopoulos, and R.A. Wentworth. Birational equivalences of vortex moduli. Topology 35 (1996), no. 3, 731–748.

[6] G.D. Daskalopoulos. The topology of the space of stable bundles on a compact Riemann surface. J. Differential Geom. 36 (1992), no. 3, 699–746.

[7] G.D. Daskalopoulos and R.A. Wentworth. Convergence properties of the Yang-Mills flow on Kähler surfaces. J. Reine Angew. Math. 575 (2004), 69–99.

[8] G.D. Daskalopoulos, R.A. Wentworth, J. Weitsman, and G. Wilkin. Morse theory and hyperkähler Kirwan surjectivity for Higgs bundles. J. Differential Geom. 87 (2011), no. 1, 81–116.

[9] S. K. Donaldson. Anti self-dual Yang-Mills connections over complex algebraic surfaces and stable vector bundles. Proc. London Math. Soc. (3) 50(1) (1985), 1–26.

[10] M.-C. Hong. Heat flow for the Yang-Mills-Higgs field and the Hermitian Yang-Mills-Higgs metric. Ann. Global Anal. Geom. 20(1) (2001), 23–46.

[11] M.-C. Hong and G. Tian. Asymptotical behaviour of the Yang-Mills flow and singular Yang-Mills connections. Math. Ann. 330(3) (2004), 441–472.

[12] A. King and P. Newstead. Moduli of Brill-Noether pairs on algebraic curves. Internat. J. Math. 6 (1995), no. 5, 733–748.

[13] F. C. Kirwan. "Cohomology of quotients in symplectic and algebraic geometry", volume 31 of *Mathematical Notes*. Princeton University Press, Princeton, NJ, 1984.

[14] S. Kobayashi. "Differential geometry of complex vector bundles", volume 15 of *Publications of the Mathematical Society of Japan*. Princeton University Press, Princeton, NJ, 1987. , Kano Memorial Lectures, 5.

[15] J. Le Potier. Systèmes cohérents et structures de niveau. Astérisque No. 214 (1993), 143 pp.

[16] I. G. Macdonald. Symmetric products of an algebraic curve. Topology 1 (1962), 319–343.

[17] J. Råde. On the Yang-Mills heat equation in two and three dimensions. J. Reine Angew. Math. 431 (1992), 123–163.

[18] N. Raghavendra and P. Vishwanath. Moduli of pairs and generalized theta divisors. Tohoku Math. J. (2) 46 (1994), no. 3, 321–340.

[19] L. Simon. Asymptotics for a class of nonlinear evolution equations, with applications to geometric problems. Ann. of Math. (2) 118(3) (1983), 525–571.

[20] M. Thaddeus. Stable pairs, linear systems and the Verlinde formula. Invent. Math. 117, no. 2 (1994), 317–353.

[21] G. Wilkin. Morse theory for the space of Higgs bundles. Comm. Anal.Geom. 16, no. 2 (2008), 283–332.

Authors' addresses:

Richard A. Wentworth
Department of Mathematics,
University of Maryland,
College Park, MD 20742, USA
raw@umd.edu

Graeme Wilkin
Department of Mathematics
University of Colorado
Boulder, CO 80309
USA
graeme.wilkin@colorado.edu

9

Manifolds with k-positive Ricci curvature

JON WOLFSON

9.1 Introduction

Let (M, g) be an n-dimensional Riemannian manifold. We say M has k-positive Ricci curvature if at each point $p \in M$ the sum of the k smallest eigenvalues of the Ricci curvature at p is positive. We say that the k-positive Ricci curvature is bounded below by α if the sum of the k smallest eigenvalues is greater than α. Note that n-positive Ricci curvature is equivalent to positive scalar curvature and one-positive Ricci curvature is equivalent to positive Ricci curvature. We first describe some basic connect sum and surgery results for k-positive Ricci curvature that are direct generalizations of the well known results for positive scalar curvature (n-positive Ricci curvature). Using these results we construct examples that motivate questions and conjectures in the cases of 2-positive and $(n - 1)$-positive Ricci curvature. In particular:

Conjecture 1 If M is a closed n-manifold that admits a metric with 2-positive Ricci curvature then the fundamental group, $\pi_1(M)$, is virtually free.

We formulate an approach to solving this conjecture based, at least philosophically, on the method used in the proof of the Bonnet–Myers theorem: A closed n-manifold that admits a metric with positive Ricci curvature (1-positive Ricci curvature) has finite fundamental group. The Bonnet-Myers theorem proves a diameter bound for a manifold with positive Ricci curvature bounded below and then uses covering spaces to conclude the result on the fundamental group. In our approach to Conjecture 1 we describe a suitable notion of "two-dimensional diameter bound", namely, the notion of *fill radius*. This idea was introduced in [G], [G-L2], [S-Y2]. We show that if a closed manifold satisfies a curvature condition that implies a fill radius bound then the fundamental group is virtually free. The proof of this statement uses covering space theory, like in the Bonnet–Myers argument, as well as some notions from geometric group theory. See [R-W] for full details and Section 4 of this chapter for an overview. However, the question: "Does a manifold with 2-positive Ricci curvature bounded below by α satisfy an upper bound on its fill radius?" remains open.

It is clear that k-positive Ricci curvature implies $(k + 1)$-positive Ricci curvature, so results on n-manifolds with n-positive Ricci curvature hold for n-manifolds with k-positive Ricci curvature, any k, $0 < k < n$. As a result of the surgery theorem we pose some interesting questions on the difference between n-positive Ricci curvature and $(n - 1)$-positive Ricci curvature.

The author was partially supported by NSF grant DMS-0604759.

9.2 Manifolds with k-positive Ricci curvature

In this section we state some structure theorems for manifolds with k-positive Ricci curvature and provide some examples. Throughout we assume that (M, g) is an n-dimensional Riemannian manifold with $n \geq 3$. Recall that n-positive Ricci curvature is positive scalar curvature and one-positive Ricci curvature is positive Ricci curvature. The following is a direct generalization of a well-known result of Gromov–Lawson [G-L1] and Schoen–Yau [S-Y1] on connect sum and surgeries of manifolds with positive scalar curvature (also see [R-S]).

Theorem 9.2.1 *Let M be a compact n-manifold with a metric of k-positive Ricci curvature, $2 \leq k \leq n$. Then any manifold obtained from M by performing surgeries in codimension q with $q \geq \max\{n + 2 - k, 3\}$ also has a metric of k-positive Ricci curvature. If M_1 and M_2 are compact n-manifolds with metrics of k-positive Ricci curvature, $2 \leq k \leq n$, then their connect sum $M_1 \# M_2$ also carries a metric with k-positive Ricci curvature.*

Proof The proof follows easily from the procedure provided in the proof in [G-L1] of the similar statement for positive scalar curvature (the case of $k = n$). We note that this procedure fails when $k = 1$ (the case of positive Ricci curvature). For the sake of completeness we will give the proof here of the connect sum result following, sometimes verbatim, the method of Gromov–Lawson. Suppose that M is an n-manifold, $n \geq 3$, and that M has a metric with Ricci curvature that is k-positive, for $2 \leq k \leq n$ (i.e., at each point the sum of any k eigenvalues of the Ricci curvature is positive). Fix a point $p \in M$ and consider a normal coordinate ball D centered at p. Following [G-L1] we will change the metric in D preserving that the Ricci curvature is k-positive such that the metric agrees with the original metric near ∂D and such that near p the metric is the standard metric on $S^{(n-1)}(\varepsilon) \times \mathbb{R}$, for any ε sufficiently small. It follows from this that 1-handles can be added and connected sums taken preserving that the Ricci curvature is k-positive, for any $2 \le k \le n$.

The method of Gromov–Lawson proceeds as follows: We consider the Riemannian product $D \times \mathbb{R}$ with coordinates (x, t), where x are normal coordinates on D. We define a hypersurface $N \subset D \times \mathbb{R}$ by the relation

$$N = \{(x, t) : (x, t) \in \gamma\}$$

where γ is a smooth curve in the (r, t)-plane that is monotonically decreasing, begins along the positive r-axis and ends as a straight line parallel to the t-axis. The metric induced on N from $D \times \mathbb{R}$ extends the metric on D near its boundary and ends with a metric on $S^{n-1}(\varepsilon) \times \mathbb{R}$. If ε is sufficiently small then by Lemma 1 of [G-L1] we can change the metric in this tubular piece to the product metric on the product of the standard ε-sphere with $\mathbb{R}$.

The key problem is to choose the curve γ so that the metric induced on N has strictly k-positive Ricci curvature at all points. To do this we begin as in [G-L1] by letting ℓ be a geodesic ray in D beginning at the origin. Then the surface $\ell \times \mathbb{R}$ is totally geodesic in $D \times \mathbb{R}$ and the normal to N along points of $N \cap (\ell \times \mathbb{R})$ lies in $\ell \times \mathbb{R}$. It follows that $\gamma_\ell = N \cap (\ell \times \mathbb{R})$ is a principal curve on N and that the associated principal curvature at a point $(r, t) \in \gamma$ is the curvature κ of γ at that point. The remaining principal curvatures at such a point are of the form $(-1/r + O(r)) \times \sin \theta$ where θ is the angle between the normal to the hypersurface and the t-axis.

Fix a point $q \in \gamma_\ell \subset N$ corresponding to a point $(r, t) \in \gamma$. Let $\{e_1, \ldots, e_n\}$ be an orthonormal basis of $T_q(N)$ such that e_1 is the tangent vector to γ_ℓ and $\{e_2, \ldots, e_n\}$ (which are tangent vectors to D) are principal vectors for the second fundamental form of N. The Gauss curvature equation relates the curvature tensor R_{ijlm} of N with the curvature tensor $\overline{R}_{ijlm}$ of $D \times \mathbb{R}$. In particular, with respect to this basis, at q:

$$\overline{R}_{ijij} = R_{ijij} - \lambda_i \lambda_j \quad \text{for } i \neq j$$
$$\overline{R}_{ijlj} = R_{ijlj} \quad \text{for } i \neq l.$$

where $\lambda_1, \ldots, \lambda_n$ are the principal curvatures corresponding to the directions $e_1, \ldots, e_n$ respectively. As above, $\lambda_1 = \kappa$, the curvature of γ_ℓ in $\ell \times \mathbb{R}$ and $\lambda_j = (-1/r + O(r)) \sin \theta$ for $j = 2, \ldots n$. Since $D \times \mathbb{R}$ has the product metric we have:

$$\overline{R}_{ijlj} = R^D_{ijlj}, \qquad\qquad \text{for } i, j, l = 2, \dots, n,$$
$$\overline{R}_{1jij} = R^D_{\frac{\partial}{\partial r}jij} \cos\theta, \qquad \text{for } i, j = 2, \dots, n,$$
$$\overline{R}_{i1j1} = R^D_{i\frac{\partial}{\partial r}j\frac{\partial}{\partial r}} \cos^2\theta, \quad \text{for } i, j = 2, \dots, n,$$

where R^D is the curvature tensor of the metric on D. It follows that the Ricci curvature of N at (x, t) with respect to the frame $\{e_1, \dots, e_n\}$ is given by:

$$\mathrm{Ric}_{11} = \mathrm{Ric}^D\left(\tfrac{\partial}{\partial r}, \tfrac{\partial}{\partial r}\right)\cos^2\theta + \kappa \sum_{j=2}^{n} \lambda_j \qquad\qquad (9.2.1)$$

$$\mathrm{Ric}_{1j} = \mathrm{Ric}^D\left(\tfrac{\partial}{\partial r}, e_j\right)\cos\theta, \qquad\qquad\qquad \text{for } j = 2, \dots, n,$$

$$\mathrm{Ric}_{ij} = \mathrm{Ric}^D_{ij} - R^D_{i\frac{\partial}{\partial r}j\frac{\partial}{\partial r}}\sin^2\theta, \qquad\qquad \text{for } i, j = 2, \dots, n, \ \ i \neq j,$$

$$\mathrm{Ric}_{ii} = \mathrm{Ric}^D_{ii} - R^D_{i\frac{\partial}{\partial r}i\frac{\partial}{\partial r}}\sin^2\theta + \lambda_i\left(\sum_{j=2, j\neq i}^{n} \lambda_j + \kappa\right), \qquad \text{for } i = 2, \dots, n,$$

where Ric^D denotes the Ricci curvature of D.

The eigenvalues of a matrix depend continuously on the entries of the matrix. Therefore for θ sufficiently small, using that $\lambda_i = (-1/r + O(r))\sin\theta$, for $i \geq 2$, the Ricci curvature at (x, t) in formula (9.2.1) is k-positive. In particular, there is an $\theta_0 > 0$ such that for $0 \leq \theta \leq \theta_0$ the Ricci curvature at (x, t) is k-positive. As in [G-L1] we bend γ to the angle θ_0 and continue γ as a straight line segment. Denote the straight line segment of γ by γ_0. Along this curve N has Ricci curvature that is k-positive. Since $\kappa \equiv 0$ along γ_0, we see that as r becomes small the Ricci curvature of N is of the form:

$$\mathrm{Ric}_{ij} = O(1), \qquad\qquad \text{for } i \neq j, \qquad\qquad (9.2.2)$$

$$\mathrm{Ric}_{ii} = \frac{(n-2)\sin^2\theta_0}{r^2} + O(1), \quad \text{for } i = 2, \dots, n,$$

$$\mathrm{Ric}_{11} = O(1).$$

Choose $r_0 > 0$ small. From (9.2.2) it follows that the the Ricci curvature of the hypersurface N has $(n-1)$ positive eigenvalues and that each of these eigenvalues is larger in absolute value than the one remaining eigenvalue (corresponding to the direction of e_1). Therefore the Ricci curvature remains k-positive and, in fact, becomes 2-positive. Consider the point $(r_0, t_0) \in \gamma_0$. Bend γ_0, beginning at this point, with the curvature function $\kappa(s)$ similar to the one used in [G-L1]:

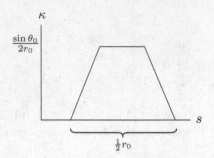

Here s is the arclength parameter. Since $\max \kappa = \frac{\sin \theta_0}{2r_0}$ we see that:

$$\lambda_i \left(\sum_{j=2, j \neq i}^{n} \lambda_j + \kappa \right) > \left| \kappa \sum_{j=2}^{n} \lambda_j \right| > 0, \quad \text{for } i = 2, \ldots, n.$$

Using (9.2.1) (and r_0 sufficiently small) this implies that along the bend the Ricci curvature of N remains 2-positive. During this bending process the curve will not cross the line $r = r_0/2$ since the length of the bend is $r_0/2$ and it begins at height r_0. The total amount of bending is:

$$\Delta \theta = \int \kappa \, ds \simeq \frac{\sin \theta_0}{4},$$

independent of r_0.

Continue the curve with a new straight line segment γ_1 at an angle $\theta_1 = \theta_0 + \Delta \theta$. Repeat the above procedure now using θ_1 where previously we used θ_0. The total bending of this procedure will then be $\frac{\sin \theta_1}{4} > \frac{\sin \theta_0}{4}$. By repeating this procedure a finite number of times (depending on $\sin \theta_0$) we can achieve a total bend of $\pi/2$. This completes the proof of the connect sum result.

For the general case of surgeries we again explain the modification of the Gromov-Lawson argument. Let $S^p \subset M$ be an embedded sphere with trivial normal bundle B of dimension $q \geq 3$. Identify B with $S^p \times \mathbb{R}^q$. Define $r : S^p \times \mathbb{R}^q \to \mathbb{R}_+$ by $r(y, x) = ||x||$, and set $S^p \times D^q(\rho) = \{(y, x) : r(y, x) \leq \rho\}$. Choose $\bar{r} > 0$ such that the normal exponential map $\exp : B \to M$ is an embedding on $S^p \times D^q(\bar{r}) \subset B$. Lift the metric on M to $S^p \times D^q(\bar{r})$ by the exponential map. Then r is the distance function to $S^p \times \{0\}$ in $S^p \times D^q(\bar{r})$, and curves of the form $\{y\} \times \ell$, where ℓ is a ray in $D^q(\bar{r})$ emanating from the origin, are geodesics in $S^p \times D^q(\bar{r})$.

We now consider hypersurfaces in the Riemannian product $(S^p \times D^q(\bar{r})) \times \mathbb{R}$ of the form:

$$N = \{(y, x, t) : (r(y, x), t) \in \gamma\}$$

where γ is, as above, a curve in the (r, t)-plane. As in the connect sum case we must choose γ so that the metric on N has k-positive Ricci curvature at all points. We first remark that $\gamma_\ell = N \cap (\{y\} \times \ell \times \mathbb{R})$ is a principal curve on N and the associated principal curvature at a point corresponding to $(r, t) \in \gamma$ is exactly the curvature κ of γ at that point.

Now fix a point $q \in \gamma_\ell \subset N$ corresponding to a point $(r, t) \in \gamma$. Let $\{e_1, \ldots, e_n\}$ be an orthonormal basis of $T_q(N)$ such that e_1 is the tangent vector to γ_ℓ, and $e_2, \ldots, e_n$ are principal vectors for the second fundamental form of N. If the metric on $S^p \times D^q(\bar{r})$ at q is the product metric then the vectors $e_2, \ldots, e_q$ can be chosen to be tangent to $\{y\} \times D^q(\bar{r})$ and the vectors $e_{q+1}, \ldots, e_n$ to be tangent to $S^p \times \{x\}$. The principal curvatures $\lambda_2, \ldots, \lambda_q$ are then of the form $(-\frac{1}{r} + O(r)) \sin \theta$, where θ is the angle between the normal to the hypersurface and the t-axis and the principal curvatures $\lambda_{q+1}, \ldots, \lambda_n$ are of the form $O(1) \sin \theta$ (i.e., are independent of r). In the general case such a simple description is not possible. However if r is small the $q - 1$ largest principal curvatures are of the form $(-\frac{c}{r} + O(1)) \sin \theta$, where c is a positive constant that can be bounded away from zero and the remaining principal curvatures are of this form $O(1) \sin \theta$. (Since the product of the principal curvatures grows like $\frac{1}{r^{q-1}}$.) We will denote the $q - 1$ largest principal curvatures by $\lambda_2, \ldots, \lambda_q$ corresponding to the directions $e_2, \ldots, e_q$ and the remainder by $\lambda_{q+1}, \ldots, \lambda_n$ corresponding to the directions $e_{q+1}, \ldots, e_n$.

The Gauss curvature equation relates the curvature tensor R_{ijlm} of N with the curvature tensor $\overline{R}_{ijlm}$ of $(S^p \times D^q(\bar{r})) \times \mathbb{R}$. In particular, with respect to this basis, at q:

$$\overline{R}_{ijij} = R_{ijij} - \lambda_i \lambda_j, \quad \text{for } i \neq j,$$
$$\overline{R}_{ijlj} = R_{ijlj}, \quad \text{for } i \neq l,$$

where $\lambda_1, \ldots, \lambda_n$ are the principal curvatures corresponding to the directions $e_1, \ldots, e_n$ respectively. As above, $\lambda_1 = \kappa$, the curvature of γ_ℓ in $\ell \times \mathbb{R}$ and $\lambda_j = (-\frac{c}{r} + O(1)) \sin \theta$ for $j = 2, \ldots, q$. The remaining principal curvatures $\lambda_{q+1}, \ldots, \lambda_n$ are of the form $O(1) \sin \theta$. For this reason, unlike in the connect sum case, they play no useful role in the following computation. Since $(S^p \times D^q(\bar{r})) \times \mathbb{R}$ has the product metric in the second factor we have:

$$\overline{R}_{ijlj} = R^{S^p \times D^q}_{ijlj}, \qquad \text{for } i, j, l = 2, \ldots, n,$$
$$\overline{R}_{1jij} = R^{S^p \times D^q}_{\frac{\partial}{\partial r} jij} \cos \theta, \quad \text{for } i, j = 2, \ldots, n,$$
$$\overline{R}_{i1j1} = R^{S^p \times D^q}_{i \frac{\partial}{\partial r} j \frac{\partial}{\partial r}} \cos^2 \theta, \quad \text{for } i, j = 2, \ldots, n,$$

where $R^{S^p \times D^q}$ is the curvature tensor of the metric on $S^p \times D^q$. It follows that the Ricci curvature of N at (y, x, t) with respect to the frame $\{e_1, \ldots, e_n\}$ is given by:

$$\text{Ric}_{11} = \text{Ric}^{S^p \times D^q}\left(\frac{\partial}{\partial r}, \frac{\partial}{\partial r}\right)\cos^2\theta + \kappa \sum_{j=2}^{n}\lambda_j \qquad (9.2.3)$$

$$\text{Ric}_{1j} = \text{Ric}^{S^p \times D^q}\left(\frac{\partial}{\partial r}, e_j\right)\cos\theta, \qquad \text{for } j = 2, \ldots, n,$$

$$\text{Ric}_{ij} = \text{Ric}_{ij}^{S^p \times D^q} - R_{i\frac{\partial}{\partial r}j\frac{\partial}{\partial r}}^{S^p \times D^q}\sin^2\theta, \qquad \text{for } i, j = 2, \ldots, n, \ \ i \neq j,$$

$$\text{Ric}_{ii} = \text{Ric}_{ii}^{S^p \times D^q} - R_{i\frac{\partial}{\partial r}i\frac{\partial}{\partial r}}^{S^p \times D^q}\sin^2\theta + \lambda_i\left(\sum_{j=2, j\neq i}^{n}\lambda_j + \kappa\right), \text{ for } i = 2, \ldots, n,$$

where $\text{Ric}^{S^p \times D^q}$ denotes the Ricci curvature of $S^p \times D^q$.

Since $\lambda_i = (-\frac{c}{r} + O(1))\sin\theta$, for $i = 2, \ldots, q$ and $\lambda_i = O(1)\sin\theta$, for $i = q + 1, \ldots, n$, if θ is sufficiently small, the Ricci curvature at (y, x, t) in formula (9.2.3) is k-positive. In particular, there is an $\theta_0 > 0$ such that for $0 \leq \theta \leq \theta_0$ the Ricci curvature at (y, x, t) is k-positive. As above we bend γ to the angle θ_0 and continue γ as a straight line segment. Denote the straight line segment of γ by γ_0. Along this curve N has Ricci curvature that is k-positive. Since $\kappa \equiv 0$ along γ_0, we see that as r becomes small the Ricci curvature of N is of the form:

$$\text{Ric}_{ij} = O(1), \qquad \text{for } i \neq j, \qquad (9.2.4)$$

$$\text{Ric}_{ii} = \frac{(q - 2)c^2 \sin^2\theta_0}{r^2} + O\left(\tfrac{1}{r}\right), \qquad \text{for } i = 2, \ldots, q,$$

$$\text{Ric}_{ii} = O\left(\tfrac{1}{r}\right), \qquad \text{for } i = q + 1, \ldots, n,$$

$$\text{Ric}_{11} = O(1).$$

(Note that since $\lambda_{q+1}, \ldots, \lambda_n$ are bounded for r small, we cannot conclude any more than $\text{Ric}_{ii} = O(\frac{1}{r})$ for $i = q + 1, \ldots, n$.) Therefore, provided $q \geq 3$, for r sufficiently small the Ricci curvature has $q - 1$ positive eigenvalues that strongly dominate all other eigenvalues. In particular, provided $k > n - q + 1$ the Ricci curvature remains k-positive. We can then bend γ to a line parallel to the t-axis and preserve k-positive Ricci curvature as above.

This construction of N determines a "tube" of k-positive Ricci curvature with two boundary components. The initial boundary component has a collar neighborhood isometric to a tubular neighborhood of S^p in M. The final boundary component has a collar neighborhood isometric to $\partial(S^p \times D^q(\varepsilon)) \times \mathbb{R} = S^p \times S^{q-1}(\varepsilon) \times \mathbb{R}$ for the product metric with the $\mathbb{R}$ factor. This allows us to glue the initial end of N to $M \setminus (S^p \times D^q(r))$ to construct a manifold M' with k-positive Ricci curvature and with one boundary component $S^p \times S^{q-1}(\varepsilon)$.

As in [G-L1] we next observe that the metric on $\partial(S^p \times D^q(\varepsilon)) = S^p \times S^{q-1}(\varepsilon)$ can be homotoped through metrics with k-positive Ricci curvature to the standard product metric of euclidean spheres on $S^p \times S^{q-1}(\varepsilon)$. The argument that accomplishes this is the same as that in Gromov–Lawson: For ε sufficiently small the metric on $\partial(S^p \times D^q(\varepsilon))$ can be homotoped through metrics with k-positive Ricci curvature to a metric on $S^p \times S^{q-1}(\varepsilon)$ that is a Riemannian submersion with totally geodesic fibers that have the euclidean metric of curvature $\frac{1}{\varepsilon^2}$. When ε sufficiently small the terms $\mathrm{Ric}(v, v)$ for vertical vectors v strongly dominate all other terms of the Ricci curvature and therefore we can deform this metric, preserving k-positive Ricci curvature (for $k > n - q + 1$), through Riemannian submersions to one with the standard metric on S^p.

Denote the induced metric on $S^p \times S^{q-1}(\varepsilon)$ by g_0 and the product of the standard metrics on $S^p \times S^{q-1}(\varepsilon)$ by g_1. The final step of the Gromov-Lawson argument shows that the homotopy constructed in the previous paragraph can be used to find a metric of k-positive Ricci curvature on $S^p \times S^{q-1}(\varepsilon) \times [0, a]$, for some $a > 0$, whose restriction to a collar neighborhood of $S^p \times S^{q-1}(\varepsilon) \times \{0\}$ is $g_0 + dt^2$ and whose restriction to a collar neighborhood of $S^p \times S^{q-1}(\varepsilon) \times \{a\}$ is $g_1 + dt^2$. We will give a proof below in Proposition 9.2.2. Using this result we can assume that the end of the manifold M' is isometric to $S^p \times S^{q-1}(\varepsilon) \times \mathbb{R}$ equipped with the product of the standard metrics on the spheres. From this the surgery result follows immediately. $\qquad \square$

To complete the proof of Theorem 9.2.1 we prove the following proposition motivated by Proposition 3.3 of [R-S], Lemma 3 of [G-L] and [Ga].

Proposition 9.2.2 *Let X be a compact n-manifold. Suppose that there is a smooth family of Riemannian metrics on X, $\{g_t\}$, $0 \leq t \leq 1$, each with k-positive Ricci curvature for some k, $2 \leq k \leq n$. Then there is a Riemannian metric g on $X \times [0, a]$, for some $a > 0$, with k-positive Ricci curvature such that the restriction of g on a collar neighborhood of $X \times \{0\}$ is $g_0 + dt^2$ and the restriction of g on a collar neighborhood of $X \times \{a\}$ is $g_1 + dt^2$.*

Proof Consider the metric $g_{f(t)} + dt^2$ on $X \times [0, a]$, where $f(t)$ is a $C^2([0, a])$ function that is monotonically increasing from 0 to 1. We will explicitly determine $f(t)$ below. We compute the curvature of $g_{f(t)} + dt^2$ at the point $(x_0, t_0) \in X \times [0, a]$. Let e_0 be the unit normal along the hypersurface $X \times \{t_0\}$ pointing in the direction $\frac{\partial}{\partial t}$. Let $\{e_1, \ldots, e_n\}$ be an orthonormal frame along $X \times \{t_0\}$ near (x_0, t_0). Then $\{e_0, \ldots, e_n\}$ is an orthonormal frame along $X \times [0, 1]$ near (x_0, t_0). Let $\{\omega_0, \ldots, \omega_n\}$ be the dual coframe. For notational convenience we use the index ranges $\alpha, \beta, \gamma = 0, 1, \ldots, n$ and $i, j, k = 1, 2, \ldots, n$. The connection one-form for the coframe $\{\omega_0, \ldots, \omega_n\}$ is $\{\omega_{\alpha\beta}\}$. The one-forms ω_{ij} depend only on the metric $g_{f(t)}$. However the one-forms ω_{0i} depend on $f'(t)$ so at (x_0, t_0) we have:

$$\omega_{0i} = O(|f'|), \tag{9.2.5}$$
$$d\omega_{0i} = O(|f''|) + O(|f'|) + O(|f'|^2).$$

The curvature two-form $\Omega_{\alpha\beta}$ on $X \times [0, a]$ is determined by the structure equation:

$$\Omega_{\alpha\beta} = d\omega_{\alpha\beta} - \sum_{\gamma} \omega_{\alpha\gamma} \wedge \omega_{\gamma\beta}. \tag{9.2.6}$$

Denote the curvature tensor of $g_{f(t)} + dt^2$ on $X \times [0, a]$ by $\bar{R}_{\alpha\beta\gamma\delta}$ and the curvature tensor of $g_{f(t_0)}$ on $X \times \{t_0\}$ by R_{ijkl}. Then by the structure equation (9.2.6) (or by the Gauss equation) at (x_0, t_0):

$$\bar{R}_{ijkl} = R_{ijkl} + O(|f'|^2). \tag{9.2.7}$$

By the structure equation and (9.2.5)

$$\bar{R}_{0jkl} = O(|f''|) + O(|f'|) + O(|f'|^2),$$
$$\bar{R}_{0j0l} = O(|f''|) + O(|f'|) + O(|f'|^2).$$

Denote the Ricci curvature of $g_{f(t)} + dt^2$ on $X \times [0, a]$ by $\overline{\text{Ric}}$ and the Ricci curvature of $g_{f(t_0)}$ on $X \times \{t_0\}$ by Ric. Then

$$\overline{\text{Ric}}_{ij} = \text{Ric}_{ij} + O(|f''|) + O(|f'|) + O(|f'|^2),$$
$$\overline{\text{Ric}}_{i0} = O(|f''|) + O(|f'|) + O(|f'|^2), \tag{9.2.8}$$
$$\overline{\text{Ric}}_{00} = O(|f''|) + O(|f'|) + O(|f'|^2).$$

Given constants, $0 < \varepsilon, \lambda \ll 1$, choose $a > 0$ sufficiently large so that there is a C^2 function $f : [0, a] \to [0, 1]$ satisfying:

(i) f is monotonically increasing,
(ii) $f(t) = 0$, for $0 \leq t \leq \varepsilon$,
(iii) $f(t) = 1$, for $a - \varepsilon \leq t \leq a$,
(iv) $|f'(t)|, |f''(t)| < \lambda$, for all t.

Using (9.2.8) since (X, g_t) has k-positive Ricci curvature for all $t \in [0, 1]$, the compactness of X and $[0, 1]$ allow that λ can be chosen sufficiently small to ensure that the metric $g_{f(t)} + dt^2$ (for f as given above) has k-positive Ricci curvature everywhere on $X \times [0, a]$. The result follows. $\square$

Let X be a compact Riemannian manifold with positive Ricci curvature. Then the manifold $X \times S^1$ has a metric of 2-positive Ricci curvature (and of non-negative Ricci curvature). Therefore, by Theorem 9.2.1, if $X_i, i = 1, \ldots, l$, are compact Riemannian $(n - 1)$-manifolds with positive Ricci curvature, the manifolds:

$$\#_i^l (X_i \times S^1), \quad l \geq 2, \tag{9.2.9}$$

admit metrics with 2-positive Ricci curvature. From this we see that compact manifolds with 2-positive Ricci curvature can have large fundamental groups. In fact the fundamental groups of the examples (9.2.9) are virtually free. This implies that the manifolds (9.2.9), when $l \geq 2$, do not admit metrics of non-negative Ricci curvature since the universal covers of (9.2.9) have infinitely many ends. In contrast the universal cover of a compact manifold with non-negative Ricci curvature splits isometrically as a product $N \times \mathbb{R}^p$, where N is a compact manifold, and therefore has one or two ends. Many other topologically distinct examples of compact manifolds that admit metrics of 2-positive Ricci curvature can be constructed by taking the connect sum of manifolds of positive Ricci curvature with the manifolds of (9.2.9).

Consider the round metric on the sphere of radius r, S_r^{n-2}, and the hyperbolic metric on the Riemann surface Σ_g of genus $g \geq 2$. When r is sufficiently small the manifolds $S_r^{n-2} \times \Sigma_g$ admit metrics of 3-positive Ricci curvature but not of 2-positive Ricci curvature. Examples of this type indicate that 3-positive Ricci curvature when $n \geq 4$ imposes much weaker restrictions on $\pi_1(M)$ than 2-positive Ricci curvature.

The surgery results of Theorem 9.2.1 do not allow any surgeries preserving 2-positive Ricci curvature (except connect sum). When $k > 2$ and $n > 3$, $q = n - 1$-surgeries preserve the curvature condition. This suggests a difference between $k = 2$ and $k > 2$ in the rigidity of the fundamental group for manifolds with k-positive Ricci curvature.

Conjecture 1 If M is a closed n-manifold that admits a metric with 2-positive Ricci curvature then its fundamental group, $\pi_1(M)$, is virtually free.

The first interesting case of this conjecture occurs when $n = 4$. When $n = 3$ the condition of positive scalar curvature is strictly weaker than 2-positive Ricci curvature. In the case of positive scalar curvature the conjecture can be answered in the affirmative. In fact, by the results of [S-Y1] and [G-L2] much more can be said about the topology of 3-manifolds of positive scalar curvature. We remark that the conjecture requires a positive curvature assumption. Requiring only 2-non-negative Ricci curvature is not sufficient. For example that manifold

$S^2 \times T^2$, where S^2 is the round 2-sphere and T^2 is the flat 2-torus, has 2-non-negative Ricci curvature and fundamental group isomorphic to $\mathbb{Z} \oplus \mathbb{Z}$.

The surgery statement of Theorem 9.2.1 shows that if $k = n - 1$ then surgery is possible provided $q \geq 3$. This is the same condition as in the positive scalar curvature case. The surgery result is used by Gromov-Lawson [G-L1] to prove that if $n \geq 5$ then every compact simply-connected non-spin n-manifold carries a metric of positive scalar curvature and by Stolz [Sz] to prove that for $n \geq 5$ every compact simply-connected spin n-manifold with vanishing α invariant carries a metric of positive scalar curvature. Gromov-Lawson prove their result for non-spin manifolds using oriented bordism. Stolz proves his result for spin manifolds using spin bordism. Since $n - 1$ positive Ricci curvature implies positive scalar curvature any necessary condition for positive scalar curvature is a necessary condition for $n - 1$ positive Ricci curvature. In light of the surgery result for $n - 1$ positive Ricci curvature exactly the same arguments apply to prove:

Theorem 9.2.3 *Let $n \geq 5$. Every compact simply-connected non-spin n-manifold carries a metric with $n - 1$ positive Ricci curvature. Every compact simply-connected spin n-manifold with vanishing α invariant carries a metric with $n - 1$ positive Ricci curvature.*

To prove the theorem, all that needs to be checked is that the generators of oriented bordism described in [G-L1] and the $\mathbb{HP}^2$ bundles used in [Sz] admit metrics with $n - 1$ positive Ricci curvature. In particular for compact simply-connected n-manifold with $n \geq 5$ there is no distinction between positive scalar curvature and $n - 1$ positive Ricci curvature.

Question 1 Is there a compact n-manifold, $n \geq 5$, that admits a metric of positive scalar curvature but does not admit a metric with $n - 1$ positive Ricci curvature?

9.3 Fill radius and an approach to Conjecture 1

In this section we will describe an approach to Conjecture 1. In the Introduction we recalled that the Bonnet-Myers theorem on the fundamental group of a closed manifold with positive Ricci curvature is proved by deriving a diameter bound on such manifolds. The 2-positive Ricci curvature condition implies a positive lower bound on the sum of any two eigenvalues of Ricci. Accordingly we seek to derive a "two-dimensional diameter bound" for such manifolds. The notion we use for "two-dimensional diameter bound" is that of *fill radius* [G], [G-L2], [S-Y2].

Let γ be a smooth simple closed curve in M which bounds a disk in M. Set $N_r(\gamma) = \{x \in M : d(x, \gamma) \leq r\}$. We define the *fill radius of* γ to be:

$$\text{fillrad}(\gamma) = \sup\{r : \text{dist}(\gamma, \partial M) > r \text{ and } \gamma \text{ does not bound a disc in } N_r(\gamma)\}$$

We say a Riemannian manifold (M, g) has its *fill radius bounded by* C if every smooth simple closed curve γ which bounds a disk in M satisfies,

$$\text{fillrad}(\gamma) \leq C.$$

Clearly if the diameter of (M, g) is bounded so is its fill radius. In particular, if for all $p \in M$ each eigenvalue $\lambda_i, i = 1, \ldots, n$ of $\text{Ric}(p)$ satisfies $\lambda_i \geq \alpha$, where α is a positive constant, then there is a constant $C = C(\alpha)$ such that the fill radius of M is bounded by C.

In [G-L2] and [S-Y2] versions of the following result on positive scalar curvature and fill radius are proved.

Theorem 9.3.1 *Let (M, g) be a complete Riemannian three dimensional manifold with bounded geometry and with positive scalar curvature S that satisfies $S \geq \alpha > 0$, for a constant α. Then if γ is a smooth simple closed curve in M which bounds a disk in M:*

$$\text{fillrad}(\gamma) \leq \sqrt{\frac{8}{3}} \frac{\pi}{\sqrt{\alpha}}.$$

The statements of the results and the details of the proofs in [G-L2] and [S-Y2] differ but the essential ideas of the proof are the same. Initially observe that if γ is a simple closed curve in M that bounds a disc Σ (more generally, Σ can be taken to be a Riemann surface with boundary) then:

$$\text{fillrad}(\gamma) \leq \sup_{x \in \Sigma} d(x, \gamma),$$

where $d(-, -)$ is the distance in M. Clearly,

$$\sup_{x \in \Sigma} d(x, \gamma) \leq \sup_{x \in \Sigma} d_\Sigma(x, \gamma),$$

where $d_\Sigma(-, -)$ denotes the distance on Σ in the induced metric. Thus the fill radius of γ can be bounded above by an upper bound on the diameter of Σ in the induced metric. For arbitrary Σ spanning γ such a bound is, of course, impossible. However in [G-L2] and [S-Y2], Σ is taken to be an area minimizer among discs spanning γ (i.e., a solution of the Plateau problem). In particular, Σ is a stable, minimal immersion. After perturbing γ inward along Σ the minimal surface is strictly stable for normal variations vanishing on the boundary. In [S-Y1] the second variation of area for a minimal surface Σ in a three manifold M is given. Let $f\nu$ be a compactly supported normal variation, where $f \in C_0^\infty(\Sigma)$ and ν is a unit normal. Then the second variation formula can be written:

$$\frac{d^2|\Sigma|}{dt^2}\Big|_{t=0} = \int_\Sigma |\nabla f|^2 + f^2 \left(K - S - \frac{1}{2}||A||^2 \right) da. \qquad (9.3.1)$$

Here $|\Sigma| = \text{Area}(\Sigma)$, S is the scalar curvature of M, K is the Gauss curvature of Σ and A is the second fundamental form. It follows then that if Σ is a strictly stable minimal surface the second order linear elliptic operator:

$$L(f) = \Delta f + f(K - S) \qquad (9.3.2)$$

is a positive operator on $f \in C_0^\infty(\Sigma)$. Denote the first eigenfunction of (9.3.2) for the Dirichlet problem by k. Then k is positive on Σ and vanishes on the boundary. It is the positivity of (9.3.2) that is used in both [G-L2] and [S-Y2] to derive a diameter bound on Σ. In [S-Y2] the argument proceeds as follows. The argument in [G-L2] is somewhat different.

Let Σ be a strictly stable minimal surface with boundary $\partial\Sigma$. Fix $x \in \Sigma$ and consider the family $\mathcal{F}$ of curves on Σ from x to $\partial\Sigma$. For σ in $\mathcal{F}$ consider the functional $F(\sigma) = \int_\sigma k(\sigma(t))|\sigma'(t)|dt$, where k is the eigenfunction of (9.3.2). Minimize F over $\mathcal{F}$. A smooth minima can be found by the direct method (as with geodesics). Denote the minimizer by τ. Since $F(\sigma)$ is invariant under reparameterization of the curve σ, we can suppose that τ is parameterized by arclength. Set the length of τ to be ℓ. Since τ is a minimizer it is stable under normal variations fixing the end points. Let μ be a unit normal vector field along τ in Σ. The normal vector field $X = \psi\mu$, where ψ is a function of the arclength parameter s that vanishes at its endpoints, is an admissible variational vector field. The second variation of the functional F at τ determines a quadratic form I given by:

$$I(\psi, \psi) =$$
$$-\int_0^\ell \left(\left(\psi'' + k^{-1}\psi'\frac{dk}{ds} + \psi(K + k^{-1}\Delta k + k^{-1}\frac{d^2k}{ds^2}) \right)\psi + \frac{2}{k^2}(\nabla k \cdot \nu)^2\psi^2 \right)k\,ds$$
$$(9.3.3)$$

where Δ is the Laplace-Beltrami operator on Σ and K is the Gauss curvature of Σ. (This expression uses that ψ vanishes at its endpoints. In the general case there is a boundary term.) Set

$$L_0(\psi) = -\left[\psi'' + k^{-1}\psi'\frac{dk}{ds} + \psi \left(K + k^{-1}\Delta k + k^{-1}\frac{d^2k}{ds^2} \right) \right]. \qquad (9.3.4)$$

Then,

$$I(\psi, \psi) = \int_0^\ell \left(L_0(\psi)\psi - \frac{2}{k^2}(\nabla k \cdot v)^2 \psi^2 \right) k\, ds \qquad (9.3.5)$$

On a minimizer $I(\psi, \psi) \geq 0$ and hence, since the term $\frac{2}{k^2}(\nabla k \cdot v)^2 \psi^2 \geq 0$,

$$\int_0^\ell L_0(\psi)\psi k\, ds \geq 0 \qquad (9.3.6)$$

for all functions ψ on $[0, \ell]$ vanishing at the endpoints.

Using that k is the first eigenfunction of L (9.3.4) becomes:

$$L_0(\psi) = -\left[\psi'' + k^{-1}\psi'\frac{dk}{ds} + \psi\left(k^{-1}L(k) + S + k^{-1}\frac{d^2k}{ds^2} \right) \right]. \qquad (9.3.7)$$

Choose a function $g \in C^\infty([0, \ell])$ that is a first eigenfunction of L_0 on $[0, \ell]$ for the Dirichlet problem. Then $g > 0$ on $(0, \ell)$, g vanishes at the endpoints and $L_0(g) \geq 0$. Hence,

$$g'' + k^{-1}g'k' + g(k^{-1}L(k) + S + k^{-1}k'') \leq 0.$$

Since $k^{-1}L(k) > 0$, this implies,

$$g^{-1}g'' + g^{-1}k^{-1}g'k' + S + k^{-1}k'' \leq 0, \qquad (9.3.8)$$

on $[0, \ell]$. Let φ be any smooth function on $[0, \ell]$ vanishing at the endpoints and multiply (9.3.8) by φ^2 to give:

$$\int_0^\ell (g^{-1}g''\varphi^2 + g^{-1}k^{-1}g'k'\varphi^2 + k^{-1}k''\varphi^2 + S\varphi^2)\, ds \leq 0$$

Integrate by parts to give,

$$\int_0^\ell \left(\tfrac{1}{2}[(g^{-1}g')^2 + (k^{-1}k')^2]\varphi^2 + \tfrac{1}{2}\left(\frac{d}{dt}\ln(gk) \right)^2 \varphi^2 + S\varphi^2 \right)ds$$

$$\leq 2\int_0^\ell \varphi\varphi'\left(\frac{d}{dt}\ln(gk) \right)ds \qquad (9.3.9)$$

Note that,

$$\left| 2\varphi\varphi'\left(\frac{d}{dt}\ln(gk) \right) \right| \leq \tfrac{4}{3}(\varphi')^2 + \tfrac{3}{4}\varphi^2\left(\frac{d}{dt}\ln(gk) \right)^2$$

$$\leq \tfrac{4}{3}(\varphi')^2 + \tfrac{1}{2}\varphi^2((g^{-1}g')^2 + (k^{-1}k')^2) + \tfrac{1}{2}\varphi^2\left(\frac{d}{dt}\ln(gk) \right)^2, \qquad (9.3.10)$$

where the second inequality follows from,

$$\tfrac{1}{4}\left(\frac{d}{dt}\ln(gk)\right)^2 \le \tfrac{1}{2}((g^{-1}g')^2 + (k^{-1}k')^2)$$

Combining (9.3.9) and (9.3.10) we have:

$$\tfrac{1}{2}\alpha \int_0^\ell \varphi^2 ds \le \tfrac{4}{3}\int_0^\ell (\varphi')^2 ds, \qquad (9.3.11)$$

where $0 < \alpha \le S$. Thus,

$$\int_0^\ell \left(-\varphi'' - \tfrac{3}{8}\alpha\varphi\right)\varphi\, ds \ge 0,$$

and so the operator:

$$-\frac{d^2}{ds^2} - \frac{3}{8}\alpha$$

has nonnegative first eigenvalue on $[0, \ell]$. Hence,

$$\ell \le \sqrt{\frac{8}{3}}\frac{\pi}{\sqrt{\alpha}}$$

This inequality holds for any $x \in \Sigma$. Therefore,

$$\sup_{x\in\Sigma} \mathrm{dist}_\Sigma(x, \partial\Sigma) \le \ell.$$

From this the theorem follows.

A fill radius bound has strong geometric implications. To illustrate this we describe two results. The first, due to [S-Y1], [G-L2], concerns closed three manifolds with positive scalar curvature. According to Milnor [Mi] any closed three manifold M has a connect sum decomposition:

$$M = S_1 \# \ldots \# S_k \#(S^2 \times S^1)\# \ldots \#(S^2 \times S^1)\# K_1 \# \ldots \# K_j$$

where the S_i are spherical space forms (this uses the solution of the Poincare conjecture) and the K_i are $K(\pi, 1)$ manifolds (a $K(\pi, 1)$ manifold is a closed manifold with contractible universal cover and fundamental group isomorphic to a group π.) Using the fill radius bound it can be shown that if, in addition, M has positive scalar curvature then no $K(\pi, 1)$ summands occur in this direct sum decomposition (see [G-L2] for a proof). In particular, this implies that if M is a closed three manifold with positive scalar curvature then the fundamental group of M is virtually free. This verifies Conjecture 1 in the three dimensional case since 2-positive Ricci curvature implies 3-positive Ricci curvature (positive scalar curvature).

The second result is due to Gromov-Lawson [G-L2]:

Theorem 9.3.2 *Let* (M, g) *be a closed n-manifold with fill radius bounded above by* β. *Then there is a distance decreasing map* $\phi : M \to \Lambda$ *onto a metric graph, such that, for each* $p \in \Lambda$,

$$\text{diameter}(\phi_{\bullet}^{-1}(p)) \leq C(\beta).$$

Proof The theorem follows from the proof of Corollary 9.11 in [G-L2]. Also see [G] Appendix 1. □

The theorem implies that closed n-manifolds with fill radius bounded above and large diameter are "long" and "thin", exactly like compact 3-manifolds with positive scalar curvature and large diameter.

We have already observed that in three dimensions 2-positive Ricci curvature implies 3-positive Ricci curvature (positive scalar curvature). Therefore closed three manifolds with 2-positive Ricci curvature bounded away from zero by α satisfy a fill radius bound depending on α. This partly motivates the following conjecture:

Conjecture 2 If M is a closed four manifold that admits a metric with 2-positive Ricci curvature bounded below by $\alpha > 0$ then M satisfies a fill radius bound depending on α.

There is another "motivation" for this conjecture: Let $(\Sigma, \partial \Sigma)$ be a stable minimally immersed surface with boundary in a Riemannian four-manifold (M, g). Using an averaging technique, the Gauss equation and an appropriate choice of variational vector field, the second variation formula for area and stability can be used to show that for any smooth function of compact support $f \in C_0^\infty(\Sigma)$ we have the inequalities:

$$\int_\Sigma \left(|\nabla f|^2 + f^2 \left(K - K_\nu - \frac{1}{2}(\text{Ric}_{11} + \text{Ric}_{22}) \right) \right) e^\sigma da \geq 0 \qquad (9.3.12)$$

and

$$\int_\Sigma \left(|\nabla f|^2 + f^2 \left(K + K_\nu - \frac{1}{2}(\text{Ric}_{11} + \text{Ric}_{22}) \right) \right) e^{-\sigma} da \geq 0. \qquad (9.3.13)$$

Here K is the Gauss curvature on Σ (in the induced metric), K_ν is the curvature of the normal bundle, Ric is the Ricci curvature on M, $\{e_1, e_2\}$ is an orthonormal frame on Σ and σ is a function on Σ that satisfies $\Delta \sigma = K_\nu$. Suppose that (M, g) has 2-positive Ricci curvature bounded below by $\alpha > 0$. Then,

$$\text{Ric}_{11} + \text{Ric}_{22} \geq \alpha.$$

In the case that the normal bundle is flat or, more generally, that the function σ has suitably small oscillation, equations (9.3.12) and (9.3.13) imply:

$$\int_\Sigma |\nabla f|^2 + f^2 \left(K - \frac{1}{2}(\mathrm{Ric}_{11} + \mathrm{Ric}_{22}) \right) da \geq 0. \qquad (9.3.14)$$

Then (9.3.14) can be used in the above argument from [S-Y2] to prove a diameter bound on Σ in the induced metric. This implies a fill radius bound and hence Conjecture 2. However, it can be shown that this line of reasoning does not, in general, hold, the obstruction being the normal curvature. This does not mean that a stable minimally immersed surface in a Riemannian four-manifold (M, g) with 2-positive Ricci curvature bounded below does not satisfy a diameter bound. Rather that the above reasoning does not apply. It remains an interesting, if unexploited, fact that the 2-positive Ricci curvature occurs in an averaged version of the second variation formula.

In the next section we will show that Conjecture 2 implies Conjecture 1 (in the four dimensional case).

9.4 The fundamental group and fill radius bounds

The main theorem on the relation between fill radius and the fundamental group can be stated:

Theorem 9.4.1 *Let M be a closed Riemannian n-manifold. Suppose that the universal cover $\pi : \tilde{M} \to M$ is given the Riemannian metric $\tilde{g}$ such that π is a local isometry. If $(\tilde{M}, \tilde{g})$ has bounded fill radius then the fundamental group of M is virtually free.*

Note, in particular, if a curvature condition implies a fill radius bound then a closed Riemannian manifold, satisfying this curvature condition, satisfies the hypotheses of the theorem. The proof of Theorem 9.4.1 is somewhat technical and is done in [R-W]. In this survey we describe the main ideas and give a complete proof under the additional hypothesis that $\pi_1(M)$ is torsion free.

One of the main ideas in the proof of Theorem 9.4.1 concerns the number of ends of a group.

Definition 9.4.2 Given a group G we define the number of ends, $e(G)$, of G to be the number of geometric ends of $\tilde{K}$, where $\tilde{K} \to K$ is a regular covering of the finite simplicial complex K by the simplicial complex $\tilde{K}$ and G is the group of covering transformation.

In particular, if G is the fundamental group of a closed manifold M then the number of ends of G is the number of ends of the universal cover $\tilde{M}$ of M. It is clear that a group G with no ends is virtually trivial and hence finite. It can be shown [E] that a group with two ends is virtually infinite cyclic. A group with three ends, in fact, has infinitely many ends. This can be seen by using the Deck transformations on the universal cover. Thus a group G can have 0,1,2 or infinitely many ends.

Our first result is on the fill radius and the number of ends of subgroups of the fundamental group. We assume that $\pi_1(M)$ is torsion free though the result holds without this assumption [R-W]. We begin with a lemma.

Lemma 9.4.3 *Let M be a closed manifold. Suppose that $N \to M$ is a covering of M such that N has fundamental group G that is finitely generated and has exactly one end. Let γ be a simple closed curve in N that represents an infinite order generator $[\gamma]$ of G. Let $\tilde{M} \to N$ be the universal cover and let $\tilde{\gamma}$ be the lift of γ to $\tilde{M}$. Then the two ends of $\tilde{\gamma}$ lie in the same end of $\tilde{M}$.*

Proof There is a finite simplicial complex K with regular covering $\tilde{K}$ such that G acts as the group of covering transformations. There is an imbedding $\iota : K \to N$ that induces an epimorphism of fundamental groups. In particular, the generators of G all lie in K. Then there is an imbedding $\tilde{\iota} : \tilde{K} \to \tilde{M}$. If $B \subset \tilde{M}$ is compact then $\tilde{\iota}^{-1}(B) \subset \tilde{K}$ is compact.

Let γ be a simple closed curve in N that represents an infinite order generator $[\gamma]$ of G. After a homotopy the lift $\tilde{\gamma}$ can be assumed to lie in $\tilde{K}$. Since G has exactly one end, any two points on $\tilde{\gamma}$, not in $\tilde{\iota}^{-1}(B)$, can be joined by a curve α in $\tilde{K} \setminus \tilde{\iota}^{-1}(B)$. The curve $\tilde{\iota}(\alpha)$ then lies in $\tilde{M} \setminus B$ and joins points on $\tilde{\gamma}$ not in B. Since this is true for any compact set B the conclusion follows. $\square$

Proposition 9.4.4 *Let M be a closed Riemannian manifold with torsion free fundamental group. Suppose that the universal cover $\tilde{M}$ of M has the property that the fill radius of every null homotopic simple closed curve is uniformly bounded above. Then no finitely generated subgroup of $\pi_1(M)$ has exactly one end.*

Proof Assume, by way of contradiction, that a finitely generated subgroup G of $\pi_1(M)$ has exactly one end. Let N be a covering of M with fundamental group $\pi_1(N)$ isomorphic to G. Since $\pi_1(M)$ is torsion free every element of G is of infinite order. Let $g \in G$ be a generator.

Denote by γ a closed minimal geodesic in N that represents g. Let $p : \tilde{M} \to N$ be the universal cover and let $\tilde{\gamma}$ be the geodesic line that is a lift to $\tilde{M}$ of γ. Let $x \in \tilde{\gamma}$ and $B_R(x) \subset \tilde{M}$ be the metric ball of radius R, center x. Then because G has exactly one end by Lemma 9.4.3 both ends of $\tilde{\gamma}$ in $\tilde{M} \setminus B_R(x)$

lie in the same end of $\tilde{M}$. The geodesic line $\tilde{\gamma}$ consists of two geodesic rays $\tilde{\gamma}_1$ and $\tilde{\gamma}_2$ beginning at x. For $i = 1, 2$, choose a point $p_i \in \tilde{M} \setminus B_R(x)$ along $\tilde{\gamma}_i$ and denote the segment of $\tilde{\gamma}_i$ from x to p_i by τ_i. Since p_1 and p_2 lie in the same end there is a curve $\beta \subset \tilde{M} \setminus B_R(x)$ joining p_1 and p_2. Denote the closed curve $\tau_1 \cup \beta \cup \tau_2$ by η. Since $\tilde{M}$ is simply connected η is null homotopic and has fill radius greater than $\frac{R}{2}$. For sufficiently large R this contradicts the fill radius bound. $\qquad \square$

To illustrate the use of Proposition 9.4.4 we prove the following special case of Theorem 9.4.1.

Theorem 9.4.5 *Let M be a closed Riemannian manifold with torsion free fundamental group. Suppose that the universal cover $\tilde{M}$ of M has uniformly bounded fill radius. Then $\pi_1(M)$ is a free group of finite rank.*

We will need the following theorem of Stallings [St]: *If G is a torsion-free, finitely generated group with infinitely many ends then G is a non-trivial free product.*

Proof Applying Stallings' theorem to $G = \pi_1(M)$, we have $G \simeq G_1 * G_2$, where each G_i is finitely generated (by Grushko's Theorem, see [Ma]). Each G_i has either two or infinitely many ends (by Theorem 9.4.4). Then apply Stallings theorem to each G_i with infinitely many ends and iterate. By Grushko's Theorem, this process terminates after finitely many steps resulting in $G \simeq G_1 * \cdots * G_k$, where each G_i is finitely generated and has two ends. Since a torsion-free, finitely generated group with two ends is infinite cyclic, we conclude that $G = \pi_1(M)$ is a free group of finite rank. $\qquad \square$

References

[E]   Epstein, D., Ends, *Topology of 3-manifolds* edited by, M.K. Fort, Jr., Prentice-Hall, 1962, 110–117.

[Ga]   Gajer, J., Riemannian metrics of positive scalar curvature on compact manifolds with boundary, Ann. Glob. Anal. Geom. **5** (1987) 179–191.

[G]   Gromov, M, Filling Riemannian manifolds, J. Diff. Geom. **18** (1983) 1–147.

[G-L1] Gromov, M, and Lawson, H., The classification of simply connected manifolds of positive scalar curvature, Ann of Math. **111** (1980) 423–434.

[G-L2] Gromov, M, and Lawson, H., Positive scalar curvature and the Dirac operator on complete Riemannian manifolds, Publ. Math de IHES, **58** (1983) 83–196.

[Ma]   Massey, W., Algebraic Topology: An Introduction, GTM 56, Springer-Verlag, New York, 1984.

[R-W] Ramachandran, M. and Wolfson, J., Fill Radius and the Fundamental Group, Journal of Topology and Analysis **2** (2010) 99–107.

[R-S] Rosenberg, J., and Stolz, S., Metrics of positive scalar curvature and connections with surgery, in Surveys on Surgery Theory, Volume 2, edited by S. Cappell, et al., Annals of Math Studies **149**, Princeton Univ. Press, Princeton, 2001.

[S-Y1] Schoen, R., and Yau, S. T., On the structure of manifolds of positive scalar curvature, Manu. Math. **28** (1979) 159–183.

[S-Y2] Schoen, R., and Yau, S. T., The existence of a black hole due to condensation of matter, Comm. Math. Phys. **90** (1983) 575–579.

[Sz] Stolz, S., Simply connected manifolds of positive scalar curvature, Ann. of Math. **136** (1992), 511–540.

[St] Stallings, J., On torsion-free groups with infinitely many ends, Ann. of Math. **88** (1968), 312–334.

Author's address:

Department of Mathematics,
Michigan State University,
East Lansing, MI 48824
USA
wolfson@math.msu.edu

Printed in the United States
by Baker & Taylor Publisher Services